JUN 1 2 2007

Einstein's Brainchild

Courtesy of California Institute of Technology Archives.

Einstein's Brainchild
Relativity Made Relatively Easy!

Barry Parker

Illustrated by Lori Scoffield-Beer

 Prometheus Books

59 John Glenn Drive
Amherst, New York 14228-2197

We gratefully acknowledge permission to reprint from *Albert Einstein: The Human Side* by Helen Dukas and Banesh Hoffman (Princeton: Princeton University Press, 1979).

Published 2007 by Prometheus Books

Inquiries should be addressed to
Prometheus Books
59 John Glenn Drive
Amherst, New York 14228–2197
VOICE: 716–691–0133, ext. 207
FAX: 716–564–2711
WWW.PROMETHEUSBOOKS.COM

11 10 09 08 07   5 4 3 2 1

Library of Congress Cataloging-in-Publication Data

Parker, Barry R.
    Einstein's brainchild : relativity made relatively easy / Barry Parker ; drawings by Lori Scoffield-Beer.
        p. cm.
    Includes bibliographical references and index.
    ISBN 978–1–57392–857–1 (hardcover)
    ISBN 978–1–59102–522–1 (paperback)
    1. Einstein, Albert, 1879–1955. 2. Relativity (Physics). 3. Physicists—Biography. I. Title.

QC16.E5.P37 2000
530'.092—dc21
[B]   00–041483

Printed in the United States of America on acid-free paper

Contents

7. TESTING THE THEORY   121

8. BLACK HOLES AND OTHER EXOTIC OBJECTS   141

9. TO THE ENDS OF THE UNIVERSE   175

List of Illustrations

Preface

A S A YOUNG MAN OF TWENTY-SIX, ALBERT Einstein produced three of the most amazing papers that the world of physics had ever seen. In the first he showed that light was not only a wave but a particle, a view that startled scientists and caused controversy for years. The second proved the existence of molecules and allowed us to calculate their size. But it was the third—special relativity—that really astounded the world. It completely changed our view of space and time. Previously they had been thought of as absolute—always the same, never changing. Einstein showed this was not true; they had a strange elasticity. In short, they were relative. And there were other results just as strange. He showed that the speed of light was the uppermost speed possible in the universe (but unattainable to us), and he showed that mass and energy were two forms of the same thing, giving a recipe for their equivalence that eventually led to the atomic and hydrogen bombs.

As hard as it is to believe, this was only the beginning. Ten years later he surpassed this amazing achievement with his general theory of relativity—a theory that is still marveled at for its elegance, beauty, and far-reaching results. It is an incomparable feat in the annals of science.

Einstein's work forms the basis of so many of the breakthroughs that have been made in cosmology, astrophysics, nuclear physics, solid state physics, space travel, and electronics that it is little wonder that his picture graces the office walls of so many scientists (and others) throughout the world, and that he was recently honored as the century's greatest thinker. With little more than a pen and pad he changed not only physics, but society, forever.

Like many geniuses, however, he was an enigma. It seemed strange to many that he never fully accepted quantum mechanics, particularly its paradoxes. The weirdness of the theory was repugnant to him. Furthermore, his stubborn thirty-year struggle to extend his general theory of relativity to include the electromagnetic field (and perhaps more) confused many who viewed it as a waste of time. But—alas!—he had the last laugh: Scientists around the world are now chasing the same rainbow he chased for so many years, and they have not yet caught it.

Even the awarding of the Nobel Prize to him in 1921 had its irony. He was awarded the prize not for his major accomplishments—special and general relativity—but for his explanation of the photoelectric effect. It was no doubt the first time the Nobel Prize was awarded for a minor achievement of a scientist, and it will likely be the last.

Einstein did not achieve his final goal, but this does not take away from his lifetime of spectacular achievements. He is still considered to be the greatest scientist of all time.

It is not possible to write the story of a scientist without occasionally using scientific terms. I have tried to explain each as I used them, but for the benefit of those new to science I have included a glossary in the back of the book.

Very large and very small numbers are a problem in a book directed at the layman. I have used scientific notation to get around writing them out. In this notation a number such as 1,000,000 is written as 10^6 (the index gives the number of zeros). Similarly small numbers such as 1/1,000,000 are written as 10^{-6}. Temperature scales are also a problem. Scientists prefer the Kelvin (K°) scale and I have used it in several places. On this scale the lowest temperature in the universe is 0° K. This is –459° F.

The line drawings were done by Lori Scoffield-Beer. I would like to thank her for an excellent job. I would also like to thank my editor, Linda Greenspan Regan, and the staff of Prometheus for their help in bringing the book to its final form. Finally, I would like to thank my wife for her support while it was being written.

1 Einstein as a Youth

E INSTEIN IS CONSIDERED TO BE ONE OF THE
greatest scientists that ever lived. He was so
famous by the time he came to America that two
Europeans, on a bet, sent him a letter addressed
"Dr. Albert Einstein, America," and it got to him at
Princeton, N.J., with only the usual delay. For a
while after he graduated, however, his future
looked anything but bright. He expected to get a
job as an assistant to one of his professors, but was
turned down by them all. For months he sent out
applications to nearby universities but no one was
interested. He finally got a job as a tutor, teaching
one student, but was fired after only a few weeks.
He was engaged to be married but his mother
hated his fiancée so much he dared not offend her
and delayed the marriage. Finally he got a position

at the patent office in Bern. He was overjoyed but within months his father died of a heart attack and he was devastated.

Despite his setbacks he continued to work on several important problems in physics and a few years later in 1905 he published five of the most important papers the world has ever seen.

Einstein was born in Ulm, Germany, on March 14, 1879, but his family moved to Munich when he was only a year old.[1] Although Jewish, he received his early education at a Catholic school. It has frequently been said that he did poorly in school, but this isn't so. He was always at the top of the class in mathematics and physical science, and although he had little interest in most other subjects and spent little time on them, he did reasonably well in them. His Uncle Jacob instilled an early interest in mathematics and a university student, Max Talmud, who dined with the family once a week, encouraged him and brought him books on math and physics.

Einstein entered the Luitpold gymnasium in 1889, the equivalent of our high school, but he soon resented the discipline and militarism of the school. One of the few highlights of this period was a geometry book that came into his hands when he was in his third year. It had a tremendous effect on him; he worked every problem in the book months before it was covered in class. With the help of Max Talmud he continued studying higher mathematics, and eventually even began studying calculus on his own.

In the early 1890s his father, Hermann, who was in a partnership with his brother Jacob, went bankrupt and decided to move to Milan, Italy, to set up a new business. Einstein was in his third year of gymnasium, and his parents decided to leave him in Munich to complete his education. He also had a military obligation to fulfill when he graduated. Left in a boarding house with a stranger, he was forlorn and lonely.

For weeks he brooded about his situation, then he began scheming as his hatred for the school increased. The only teacher he liked was his math teacher, and in return most of his teachers had a negative opinion of him, not because he was a poor student, but because he was rebellious and moody. He finally went to his doctor, who was a friend of the family, and explained the situation

Fig. 1.1.
Einstein with
his sister Maja.
He is about
five years old.
Courtesy of
Lotte Jacobi.

to him. He asked the doctor to write a letter saying he was on the verge of a nervous breakdown and needed the climate of Northern Italy to recover (this is, of course, where his parents were). The doctor reluctantly agreed. Einstein also got a letter from his math teacher stating that his mathematical development was equal to that of a gymnasium graduate. Ironically, as he was getting ready to leave, his homeroom teacher, who was also his Greek teacher, called him in and told him he was a distraction in the classroom, and it was best he leave. The same teacher had told his father earlier that he "would never amount to anything."

Days later Einstein was on the train to Milan. He hadn't told his parents he was coming and was worried about their reaction. As expected they were surprised to see him, and disappointed that he had dropped out of school. To placate them he promised to take the entrance exams at the Polytechnic in Zurich the following autumn. Graduation from a high school was not necessary if you passed these exams. But when his mother inquired, she found that the minimum age for taking the exams was eighteen; Einstein was only fifteen, and would only be sixteen when they were given. With the help of a friend in Zurich, Einstein's mother made an appointment with one of the councilors at the Polytechnic and managed to convince him to let Einstein take the exams.

Over the next few months Einstein studied for the exams, but they would cover most high school subjects and Einstein hated studying botany, zoology, French, history, and so on. Despite reprimands from his mother he spent most of his time studying math and physics, subjects in which he had already demonstrated great strength. His study of physics, in fact, extended well beyond the usual high school curriculum. He was particularly interested in the relation between electric and magnetic fields, and was fascinated by electromagnetic waves, the waves that are all around us in the form of radio and TV waves. They span a spectrum from very long radio waves all the way to extremely short x and gamma rays, with ordinary light in between.

The discovery that light and other electromagnetic radiations were waves posed a dilemma to scientists in the mid-1800s. Waves required a medium of some sort to propagate them. If you threw a stone into a lake a wave moved out from the position where it struck the water, but if you took away the water (the lake) there would be no wave. Waves obviously require a propagating medium so scientists invented one. They called it the ether; it was a strange substance that permeated the entire universe, strange because it was invisible, had no taste or smell, and yet it had to be rigid enough to propagate waves. Furthermore, there seemed to be no way of getting rid of it. If you evacuated a bell jar (in other words, you took all the air out of it), light would still pass through it; therefore the ether was still there.

The concept was so strange that many people found it hard to accept, but there seemed to be no way around it. How could light—a wave—propagate if there wasn't a propagating medium? Einstein was only fifteen when he began thinking about the problem. He would eventually grow to dislike the idea of an ether, but in 1895 he was still intrigued with it. He wrote a five page essay on the subject expressing his opinions, and sent it to his Uncle Caesar in Stuttgart. In the essay he proposed an experiment to determine if electricity, magnetism, and the ether were connected. Caesar was so impressed he wrote Einstein's mother telling her that he was sure Einstein had a great future ahead of him.[2]

It was about this time that Einstein began to consider what it would be like to ride a light beam through space. He tried to visualize what things would look like from the beam. It was obvious that a clock sitting on the beam with him would run as usual, but for someone looking at the clock from Earth no time would appear to pass. The hand of the clock would always be in the same position. The question continued to haunt him for years (see figure 1.2).

In the fall of 1895 Einstein headed for Zurich to take the entrance exams at the Polytechnic. He was well-prepared in math and physics but unsure of himself in most other subjects, and as expected he did extremely well in math and physics and poorly in most other subjects. Overall his grades were not good enough for entrance to the Polytechnic, but officials were so impressed with his math and physics grades that they encouraged him to complete his last year of high school at one of the nearby towns and enter the following year. He would not have to retake the exam. The pro-

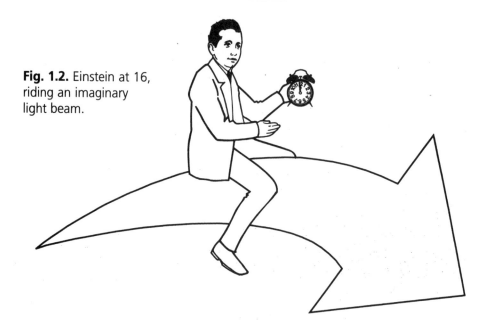

Fig. 1.2. Einstein at 16, riding an imaginary light beam.

fessor who marked the physics section of the exam was so impressed that he invited Einstein to audit his lectures. Einstein had to decline because he would not be staying in Zurich.

A school in Aarau, a small town about twenty-five miles from Zurich, was selected. Einstein stayed with the Wintelers; Josh Winteler was one of the teachers at the school. Einstein's fear that the school might be like the one he had hated so much in Munich was soon quelled. He found it was much more open and less concerned with discipline. Moreover, with the Winteler family, he could speak his mind. He soon looked upon them as his second family, and for a while he was romantically involved with one of the Winteler girls, Marie. Einstein did much better at this school, and got along well with the teachers. Just prior to graduation he was selected to go on a field trip to one of the mountains in Switzerland, a fateful trip in which he almost fell to his death.

Einstein graduated that summer and the following fall he went to the Polytechnic in Zurich. Because of his hatred of Germany he had renounced his German citizenship and was now stateless. He

Fig. 1.3.
Einstein as a
student at the
Zurich Polytechnic
Institute.
Courtesy of
Lotte Jacobi.

remained stateless for the next four years until he became a Swiss
citizen. The course he signed up for would allow him to teach
math and physics at the high school level. There were five people
in his class—three majoring in math and two in physics. One of the
math majors, Marcel Grossman, became a close friend and the
other physics major, Mileva Marić, later became his wife. It might
be thought that someone who would eventually change the world

of science so dramatically would immediately be at the top of his class at the Polytechnic. After all, he only had to take mathematics and physics now, without the distraction of other subjects that he had no interest in. But again there were problems, which for the most part could be traced to his rejection of authority, his reluctance to accept anything without proving it for himself, and his wide-ranging interests. While he was fascinated with the discoveries being made in physics, he was easily bored by routine classwork. His interests extended beyond the curriculum, and he spent most of his time studying on his own; as a consequence he skipped many classes. In addition, he enjoyed working in the laboratory and spent more time in it than was required. For most this would have been a risky course to take, but his friend Marcel Grossman took excellent notes, and he loaned them to Einstein regularly. (His notes were so good, that they are still on display at the university.) It was these notes that enabled Einstein to pass.

With his cocky, know-it-all attitude, Einstein was not a favorite among his teachers. His physics teacher, Heinrich Weber, the one who invited him to audit his lectures after he failed the entrance exams, was now beginning to dislike him, and his math teacher, Herman Minkowski, called him a "lazy dog."

Luckily for Einstein there were only two exams that had to be passed: the intermediates, taken after the second year, and the finals, taken at the end of the fourth year. Einstein's capabilities were evident in the intermediates. Despite skipping many classes, with some quick cramming he placed at the top of his class. He was slightly embarrassed that he had beaten Grossman after borrowing his notes.

Einstein's neglect of his classes wasn't a result of disinterest or laziness. During most of his college years his mind was whirling with activity. He was studying books on his own by all the top scientists of the day: Hermann von Helmholtz, Ludwig Boltzmann, Gustav Kirchhoff, Ernst Mach, and James Maxwell. He was particularly impressed with Maxwell's electromagnetic theory. To his dismay Weber did not cover the theory in class; when Einstein asked him why, Weber told him he did not feel it was part of the accepted curriculum. Einstein never forgave him.

During this time Einstein's interest in the ether never wavered, and when he was in his fourth year he thought up an experiment that would allow it to be tested through the use of small mirrors and thermocouples (devices for measuring temperature). He went to his teacher, Weber, asking if he could perform the experiment, but Weber turned him down, telling him it would be a waste of time. Interestingly, the experiment was similar (but much less sophisticated) to one performed earlier by Michelson and Morley. Weber also forced Einstein to rewrite his senior thesis on regulation paper just before the finals, denying him valuable study time.

Einstein studied with Mileva for the finals, again using Grossman's notes extensively. This time, however, he didn't place at the top of his class. In fact, all three mathematicians beat him. He was fourth, beating only Mileva, who failed. His average was 4.91 out of 6; Mileva got 4.0.[3]

Exams . . . Ugh!

Comment by Einstein on his final exams at the Swiss Polytechnic in Zurich:

"For the exams, one had to stuff oneself with all this rubbish, whether one wanted to or not. This conclusion had such a negative effect on me that after my finals, the consideration of any scientific problem was distasteful for me for a whole year."[4]

Einstein, now a graduate of the Polytechnic, became engaged to Mileva, but when his mother found out she was Serbian, she was enraged. At one point she flung herself on a bed weeping, begging him not to marry her. He dared not go against her wishes and postponed the marriage.

After graduation he expected to get a job as an assistant to one of his professors. Assistants were needed in the labs. As the only physicist (besides Mileva, but she had failed), he expected Weber to give him a job, but Weber selected two mechanical engineers with little experience in laboratory physics. He tried most of the other professors, but to no avail. Finally he began applying to other universities, but it didn't help. And as the months passed, he grew more and more depressed. In the meantime he continued to fight with his mother over Mileva; she was still determined to stop the marriage.

Einstein finally went back to Zurich to work on a doctoral thesis and to be with Mileva. He had originally thought about working under Weber, but changed his mind and went to the University of Zurich where he talked to the physics professor, Dr. Alfred Kleiner. Kleiner accepted him as a doctoral student. Interestingly, Einstein submitted what would eventually become his special theory of relativity to Kleiner, but it was rejected because no one at the university could understand it. At this stage it was far from complete, however.

Einstein's interests now turned to capillarity (the rise of water and other fluids in tubes of small diameter). He submitted a paper on the subject to *Allenen Der Physiks*, the most prestigious physics journal in Europe, and was overjoyed when it was accepted. He was now a published scientist and hoped the publication would help him get a job, but it did little. He continued working on capillarity and soon sent off another paper for publication. It was also accepted.

The one ray of light in his life at this time was a letter from the Immigration Department that he had been accepted for Swiss citizenship. He had been interviewed earlier, and a detective had checked him out. He was happy, but knew he would now have to serve in the Swiss military. Strangely, even though he had rejected his German citizenship to avoid serving in the German military, he now looked forward to serving in the Swiss army. He eagerly reported for the medical exam and was surprised when he was rejected because of flat feet and varicose veins. (They would make it difficult for him to stand the rigors of the extensive marching that would be required.)

He stayed on in Zurich with Mileva, but with no job his money eventually gave out and he had to return to Milan to live with his family. His father's business was now on the verge of bankruptcy so his family could give him little help. Einstein continued sending letters of application from Milan; letters went to the Universities of Göttingen, Stuttgart, Vienna, Bologna, and Pisa. But nobody was interested. He could not understand why he was being so completely ignored; finally he came to the conclusion it had to be Weber. He was using him as a reference and was sure Weber was writing poor letters of reference. He therefore stopped using him.

Soon after reprints of his publication on capillarity were sent to him, he mailed one to Wilhelm Ostwald at Leipzig University. Ostwald was a world expert on capillarity and Einstein was sure he would be impressed, but Ostwald never answered him. His father felt so sorry for him that he decided to write Oswald, hoping it might help. Part of his letter said:

> I shall start by telling you that my son Albert is 22 years old, that he has studied at the Zurich Polytechnic for 4 years, and that he passed his diploma examinations in mathematics and physics with flying colors last summer. Since then, he has been trying unsuccessfully to obtain a position as an Assistant. . . .
>
> My son feels profoundly unhappy with his present lack of position, and his idea that he has gone off the tracks with his career and is now out of touch gets more and more entrenched every day.[5]

He ended the letter begging him to offer Albert an assistantship, or at least write him a few words of encouragement "so he might recover his joy in living and working."

Nothing came of the letter.

Einstein had reluctantly mentioned his problems to his good friend Marcel Grossman. Grossman talked to his father who, in turn, talked to a good friend, Friedrich Haller, the director of the patent office in Bern. As it turned out a position was opening up in the patent office, and although they usually hired engineers, Einstein's background impressed him. The position would be advertised in the paper soon, and Einstein was encouraged to apply. It would, however, be months before the job became available.

Einstein applied for the position and felt confident he would have a good chance of being hired. But he was so broke that he needed something to see him through the next few months. He finally managed to get a temporary job teaching at the small town of Winterthur as a replacement for one of the teachers who was on military duty, but it lasted only three months. As the job was ending he discovered Mileva was pregnant; there was no way, however, that he could marry her at this point, and she went home to her parents in Yugoslavia to have the baby.

Einstein got another temporary job in Schlaffhaussen, a nearby village, at the institute of a Dr. Neusch. He would be tutoring one student in math. Einstein disliked Neusch from the beginning, finding him overbearing and militaristic. What made things worse was that Einstein had to stay with the Neusches and eat his meals with the family. It was a trying experience and only a matter of time before there was conflict. Einstein finally told him off and was fired on the spot.

It would still be several months before the job in the patent office in Bern would be available—if he got it. He had little money and wasn't sure what to do. He had submitted a doctoral thesis to Kleiner at the University of Zurich, so he went to Zurich to check if it had been accepted. It had not, so he withdrew it, partly to get his doctoral deposit of 220 francs back. He then went to Bern to wait for the job in the patent office.

He knew his money wouldn't last long, so he placed an ad in the Bern paper offering tutoring in math and physics. Within a short time he had two students: Maurice Solovine and Conrad Habicht. He had known Habicht earlier. Einstein had never been one for standing before a group and lecturing, and soon the tutoring sessions became discussion sessions with Einstein as group leader. Einstein, in fact, preferred to do the discussing as he walked, so many of their sessions took place with the three men on hiking trails. On one occasion they hiked to the peak of a nearby mountain during the night and didn't get back until the following morning.

Their discussion covered many areas: electricity, magnetism, dynamics, thermodynamics and even philosophy and logic. They studied books by Jules-Henri Poincaré, Ernst Mach, Karl Pearson, John Stuart Mill, and others. The money that Einstein brought in from the group, however, was small and he was always on the verge of being broke. The job at the patent office was still weeks away, but Einstein waited patiently and ate sparingly. The meeting continued regularly, and eventually the group began to call themselves the "Olympic Academy." It was a pretentious sounding name, but the three men never took themselves seriously; indeed, they loved to play practical jokes on one another. Einstein thoroughly enjoyed the discussions and there's no doubt that they helped develop his ideas. Later in life he said that it was more of an "academy" than some of the ones he was later associated with.

Einstein was finally called in to the patent office for an interview by Haller. He was questioned extensively, then a few days later he was hired as Technical Expert Class III with a salary of 3,500 francs a year. It was more money than he had seen in a long time. He wrote Mileva telling her the good news, but mentioned nothing about a wedding. Her baby had been born by now and she was waiting word from him. Within a short time Einstein moved to a better apartment, but he had barely settled in when he received a letter from home. There was an emergency: his father had had a serious heart attack. Einstein rushed home and within few days his father died. He had always been close to him and was devastated. It took him years to get over it.

His job at the patent office was examining inventions to see if they were worthy of a patent, and that they were not infringing on any other patents. His work schedule left him with a considerable amount of spare time in which he was able to work on his own projects in the evenings. Several problems interested him. One was the problem of absolute motion, in other words, motion that was independent of a reference system. He was particularly interested in its relation to electric and magnetic fields. Newton had postulated that absolute motion existed; he couldn't explain why, but he was convinced that space itself could be regarded as a fixed frame of reference. Stars and galaxies would then have motion relative to space. To Newton time was also absolute; in other words, it was the same for all observers throughout the universe. But Einstein was not convinced. Something seemed to be wrong.

Einstein was also interested in proving the existence of atoms.

Fig. 1.4. Einstein in 1905, the year he published his special theory of relativity. Courtesy of Lotte Jacobi.

As strange as it might seem, a number of well-known scientists, Ernst Mach and Friedrich Ostwald amongst them, still did not believe in the existence of atoms. Einstein decided it was time to prove once and for all that they existed, but he would have to do it theoretically. It would make a good problem for his doctoral thesis. He was also interested in Max Planck's new concept called "quanta," but it was too controversial for a thesis.

Soon after marrying Mileva, Einstein applied for a teaching position called a "privatdozent" at the University of Bern. It was an unpaid position, with fees being paid by the students, but it was a prerequisite to a professorship. He did not have a doctorate, but had published several papers and hoped they would be sufficient, but they weren't. He was turned down.

The "Olympic Academy" continued to flourish and the discussions helped Einstein see his way through the tangle of new ideas he was considering. His head was now whirling with ideas, and the time was ripe for their development. But alas, Solovine finally got a job teaching at a nearby town and left, and then a few months later Habicht left and the Academy dissolved. Einstein felt the loss; he needed people to talk to about his ideas. Then one day a notice appeared on the bulletin board at the patent office. A position was available. Einstein thought of his close friend Michele Besso, who was an engineer and would be ideal for the job. Furthermore, he had always taken an interest in Einstein's work. He would be an ideal sounding-board, and Einstein was also sure he would be helpful in other ways. He informed Besso of the position, and indeed Besso got it. To Einstein's chagrin he was actually taken on at a higher wage than Einstein.

The two men talked extensively about physics as they walked to and from work, and they frequently got together in the evenings. One of the problems Einstein was now interested in was "the electrodynamics of moving bodies," and it would soon form the basis of his special theory of relativity.

His First Honorary Doctorate

While still in the patent office in Bern, Einstein received a large envelope in the mail. Opening it he found a letter in fancy, colorful script. Thinking it was an advertisement of no importance he threw it in the wastebasket. Only later did he learn that it was the announcement that he, along with Marie Curie and Wilhelm Ostwald, would be awarded honorary doctorates by Geneva University. When university officials received no reply they asked one of his friends to check. The friend persuaded Einstein to go to Geneva without telling him why. Einstein was quite unprepared and surprised when he found out he was to be awarded an honorary degree.

The Michelson-Morley Experiment

R ELATIVE AND ABSOLUTE MOTION ARE TWO OF the most fundamental concepts in physics. You are no doubt familiar with relative motion. When a car passes the one you are in, it is moving relative to you. Indeed, all motion on Earth is relative to the earth's surface. An airplane flying overhead is moving relative to the earth. Absolute motion, on the other hand, is more difficult to visualize. Suppose, though, that we wanted to determine our absolute motion in the universe. How would we go about it? First of all we would have to determine every type of motion the earth is undergoing. It is, of course, spinning on its axis and revolving around the sun, but the sun, in turn, is orbiting our galaxy, the Milky Way. Furthermore, the sun is being pulled slightly by nearby stars so it also has a small drift as it orbits. And finally, our galaxy also has a motion through space; as we will see later, all galaxies are expanding away from one another. With so many independent motions, the earth's true or absolute motion through the universe, which is a result of all its motions, is obviously quite complex.

Newton didn't know about galaxies or the expansion of the universe; nevertheless, he was worried about the concept of absolute motion. It had important implications for his laws of motion, particularly his first law which can be stated as follows: Every object in the universe moves uniformly in a straight line unless acted on by a force (a push or a pull). Newton wasn't sure how to determine absolute motion but he was convinced it existed. He did know that some kind of frame of reference was needed, and since one didn't seem to be available he assumed he could use space itself. He was troubled with the idea, but clung to it nevertheless. To him all motion relative to what he referred to as "physical space" was absolute. Scientists soon realized, however, that this didn't make sense.

Most of the time we have little trouble identifying relative motion, but there are times when things can be confusing. Assume, for example, that we are on a high-speed monorail that is so smooth and free of vibrations that it can start off without us knowing. Assume further that all the blinds are down on one side of the monorail, and on a track on the other side of it there is another monorail. Looking out we see that it is sitting still; in other words, it is motionless. But is it? As I said earlier our monorail is so smooth it can get under way without us knowing it. This means that both monorails could be traveling fifty miles an hour side-by-side, or they both could be sitting still. The only way you could know for sure is by going to the opposite side of the monorail, lifting a blind, and looking out.

As it turns out there are an infinite number of possibilities with two monorails that are side-by-side. Assume again that you are in one of them, and that the second one appears to be passing you at 10 miles an hour. In reality it could be standing still and you could be traveling 10 miles an hour in the opposite direction, or your monorail could be going 50 miles an hour and it could be going 60 miles an hour, and again the only way you would know for sure which is the case is by looking out the window on the opposite side at some reference point.

If we can't use space itself, as Newton suggested, can we use anything else in the universe as a frame of reference? We obviously can't use galaxies since we don't know how fast they are going, and there doesn't appear to be anything else available. But scientists in the nineteenth century concluded that there was, indeed, something we could use.

THE ETHER

As I mentioned earlier, after it was discovered that light was a wave, scientists had to invent a medium to propagate the wave. This medium was called the ether, and it had some particularly strange properties: it filled the universe and it penetrated transparent solids such as glass. Furthermore, it didn't impede objects such as stars and planets that moved through the universe, so it had to be tenuous, like a gas. Yet it had to vibrate rapidly to transmit light waves, so it had to be rigid. How could it be both tenuous and rigid? This seemed to be a contradiction. And there were other problems. It's no wonder the ether was perplexing. It had to exist, yet it didn't make sense.

One of its more acceptable properties was that it had to be immobile. In essence the ether had to be at rest, which pleased scientists. They therefore had something they could use as a frame of reference. The ether filled the universe and was at rest in it; therefore, any motion through it was absolute. But how could we determine this motion? The best way was to use a beam of light. Light traveled through the ether; in fact, it was just a vibration of the ether. Furthermore, the velocity of light through the ether was well known; it was 186,172 miles per second. (Early on it wasn't known with high accuracy, but we won't worry about that for now, as it has no bearing on the argument.) I should also mention that the velocity of light is less when it passes through transparent materials such as glass or water.

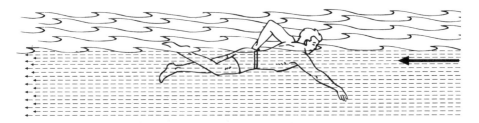

Fig. 2.1. Swimmer swimming against a current in a river.
If he swims at 5 mph and the current is 3 mph against him,
his speed relative to the shore will be 2 mph.

To understand the ether better, let's assume it is a sea (or a lake, if you like). If we create a wave on this lake by throwing something into it, we see the wave move away from the spot where the object struck the water. Let's assume it expands outward at 10 miles an hour. We could obviously chase the wave. If we set our boat going 5 miles an hour in the direction the wave was traveling, it would only be going 5 miles an hour faster than us. Indeed, if we traveled at 10 miles an hour we would keep up with it. Furthermore, if we traveled at 5 miles an hour in the opposite direction it would appear to be going 15 miles an hour relative to us. All this is pretty straightforward.

Now, let's consider a light beam in our sea of ether. It is like our water wave, but its speed is 186,172 miles per second (or 299,795 kilometers per second). Assume that you are sitting still in the ether and that you turn on a spotlight. The beam from the spotlight will move out from you at 186,172 miles per second. It is important to remember that it is the ether that is carrying the light beam. With this in mind, let's chase the light beam. Suppose we jump in a rocketship and travel at 86,172 miles per second (see figure 2.2). The light beam should therefore be going 100,000 miles per second faster than us. Indeed, let's increase the rocketship's speed to 186,172 miles per second. It is reasonable to assume that we would then be keeping up with the light beam. As we saw in chapter 1, Einstein tried to imagine what a trip like this would be like (i.e., riding a lightbeam) when he was sixteen, and it worried him.

With this in mind, let's imagine an experiment that would allow us to determine absolute motion. Consider the ether from our vantage point here on Earth. It is presumably fixed to the universe as a whole, but we know that the earth is moving around the sun, so we are therefore moving through the ether. The earth moves around the sun at a speed of about 18.5 miles a second, so the earth is moving through the ether with this speed. This means that from the earth we should feel an ether wind of 18.5 miles a second. You have a similar situation in a moving car.

Assume there is no wind and you jump into your car and accelerate to 60 miles an hour. If you put your hand out the window you feel a breeze of 60 miles an hour hitting it. In the same way we would presumably feel an "ether wind" from earth because of the earth's velocity. Actually, there's a problem I haven't mentioned that you may have noticed. We have assumed that the sun is at rest relative to the ether, and it might not be. We can take care of this easily enough, however. The earth goes around the sun, so by measuring the ether wind on earth at various times during the year we could easily determine the ether wind at the sun and eliminate it from our calculations. So for now we'll ignore it.

In theory, then, it seemed that we could perform an experiment

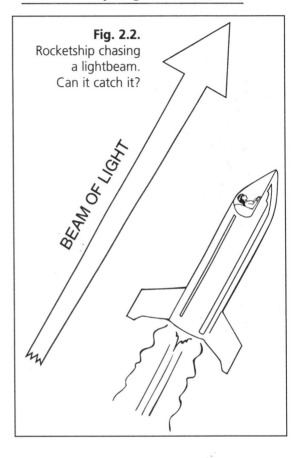

Fig. 2.2.
Rocketship chasing a lightbeam. Can it catch it?

BEAM OF LIGHT

that would allow us to determine the velocity of the ether wind on Earth, and that in turn would allow us to determine our absolute motion through space. The mathematical physicist Clerk Maxwell, the discoverer of electromagnetism, suggested the experiment in 1875, but there didn't appear to be an easy way of doing it.

MICHELSON

A physics professor in Annapolis, Maryland, Albert Michelson heard of Maxwell's suggestion about 1880. Born in Prussia in December of 1852, Michelson was only four years old when his parents brought him to the United States. They soon moved to the west, settling in Nevada, and when Albert was sixteen he decided he wanted to go to the Naval Academy in Annapolis. He was highly intelligent and graduated near the top of his class; his seamanship, however, did not match his academics and when he graduated in 1873 he began teaching in the science department at Annapolis. His interests soon began to center on light and optics. The speed of light had been determined in 1670 by the Danish astronomer Olaus Roemer using eclipses of the moons of Jupiter, but it was an approximate value. In 1849, however, the French physicist Armand Fizeau arrived at a more accurate value using a cogged wheel, and it was improved by Fizeau and a colleague, Jean Foucault, a few years later.

Michelson soon became interested in obtaining an even better value for the speed of light, and to his surprise he managed to do it with a relatively crude apparatus. Encouraged by his success, he began to dream about much more ambitious experiments, but he soon realized his knowledge of optics was limited. He therefore resigned from the navy and went to Europe to expand his horizons. He studied in both France and Germany. In Germany he worked in the laboratory of the renowned physicist, Hermann Ludwig Helmholtz. The phenomenon known as interference had been discovered about fifty years earlier and there was still considerable interest in it. Lines of interference (fringes) could be created by superimposing two light beams that were out of phase. If two light waves were exactly out of phase (and identical otherwise), they would cancel one another and leave darkness (see figure 2.3). On the other hand, if two waves were in phase (or nearly so), they would reinforce one another and the intensity of the light would increase (see figure 2.4). In a general case when you brought two waves together you would get a series of bright and dark lines, referred to as interference fringes. The dark lines occur where the waves destroy one another, and the bright ones where they reinforce one another (see figure 2.5).

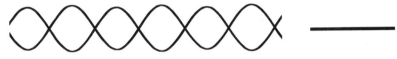

Fig. 2.3. Waves out of phase and exhibiting destructive interference.

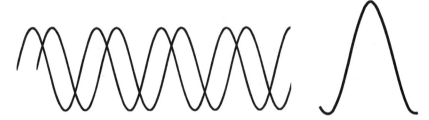

Fig. 2.4. Waves approximately in phase so they will enhance one another with constructive interference.

Fig. 2.5. Interference fringes.

Michelson discovered an ingenious way of splitting a beam of light into two beams and bringing them back together again. If they were out of phase when brought back together, interference fringes would occur. He used a semi-silvered mirror, one that would let half the light through and reflect the other half. If you held this mirror up to the light you would be able to see through it but you would also see that it was silvered and therefore partially reflective. Using other mirrors Michelson was able to bring the two beams back together again so they created interference fringes.

It was soon obvious to Michelson that he could use his technique to perform the experiment that Maxwell had suggested. Using the semi-silvered mirror he could send half of the light at right angles to the other half. If we now assume for simplicity that the direction of the ether wind on Earth is in the direction of the original beam, we will have a beam traveling parallel to the ether wind and one traveling perpendicular to it. If mirrors are now set up so these two waves are reflected back and superimposed, we will get interference fringes (like those in figure 2.5). These interference fringes give us a measure of the velocity of the ether wind, which in turn would give us our absolute velocity.

Let's look at this a little closer. Consider the earth as it goes around the sun in its orbit (see figure 2.6). Assume we project a light beam in the direction of the earth's velocity (position A). It will go out at 186,172 miles per second relative to the fixed ether. But the earth has a velocity of 18.5 miles per second in orbit, and it is therefore catching up with the beam. This means the beam should appear to be going 186,172 – 18.5 = 186,153.5 miles per second. If we wait six months and try the experiment at B, again projecting the light ray out in the same direction, we will have to add the two velocities to get the velocity of light relative to us. There isn't anything strange about this; after all if you threw a baseball at 30 miles an hour out the window of a car that was traveling at 60 miles an hour, making sure it was projected in the same direction that the car was traveling, the baseball would have a speed of 90 miles an hour relative to the ground. I don't think anyone would argue with that.

Michelson set up his experiment in Helmholtz's laboratory in Berlin in 1881. One of the light beams in his interferometer was in the direction of the earth's travel and he expected the earth to catch up with it slightly. His experiment was rather crude in that few precautions had been taken. Experiments of this type are susceptible to many types of errors. Tiny vibrations, hundreds of feet away, for example, can completely throw it off. Michelson didn't worry about

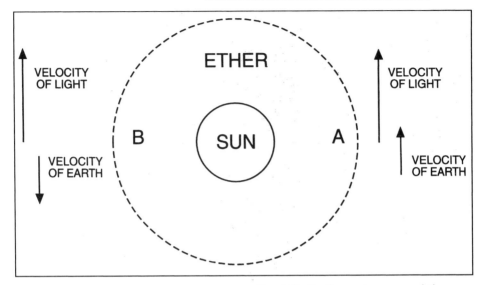

Fig. 2.6. Earth moving around the sun.
At position A light beam and Earth are traveling in the same direction.
At position B they are in opposite directions.

this because he knew his interferometer was easily sensitive enough to measure the difference between a velocity of 186,172 miles per second and 186,153.5 miles per second. He performed the experiment expecting the latter velocity and to his surprise he got the former. He repeated the experiment several times but always got the same result. He was disappointed and thoroughly confused, but he knew he hadn't taken extreme precautions; he would have to redo the experiment, making sure there were no errors.[1]

THE MICHELSON-MORLEY EXPERIMENT

Michelson returned to America and took a job at the Case School of Applied Science (now the Case Western Reserve University) in Cleveland, Ohio. Though he had to teach, he had considerable time for research, and the first thing he planned on doing was the experiment he had performed earlier in Berlin. I should mention, inciden-

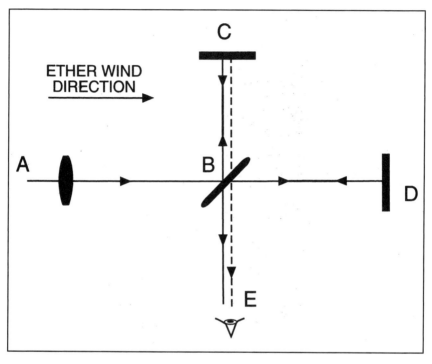

Fig. 2.7. The Michelson-Morley Experiment. Light from a source A is directed at a semi-silvered mirror at B. Half the light is reflected at B to C, half is transmitted to D. Mirrors at C and D reflect the light back toward B. Half the light from C passes through the mirror and travels to E; also half the reflected light from D is reflected at B and travels to E. The two beams are traveling parallel from B to E and interfere with one another. Interference fringes are therefore observed at E. If we assume the direction of the ether wind is ABD then the beam from B→C→B actually goes as ↗↘. The beam from B→D→B goes as →← so there is a difference in pathlength. When they reform they will therefore be out of phase and interference fringes will occur. By measuring these interference fringes Michelson could determine the velocity of the ether wind.

Fig. 2.8. The Michelson-Morley apparatus.

tally, that this is the experiment that Einstein wanted to do when he was in his fourth years at the Polytechnic. His experiment was much less sophisticated than Michelson's and it's unlikely he would have made the discovery Michelson did, even if he had performed it. Furthermore, Michelson had actually performed the experiment by the time Einstein wanted to do it; Einstein just hadn't heard of it.

As Michelson made plans to redo the experiment in Cleveland, he became friends with a chemist at nearby Western Reserve University by the name of Edward Morley. Morley was interested in light and was keen on working with Michelson. Together they decided to do the experiment in a laboratory in Morley's basement under very precise conditions. They would make sure there was no chance for error. They began by mounting the apparatus on bedrock; the base of the interferometer was then made of stone one foot thick and placed in liquid mercury so that the entire apparatus could be swung around in various directions (see figure 2.8). There was literally no chance of external vibrations affecting it now.

By late 1886 everything was ready to go. They redid the experiment—but there was still no change. The earth was not catching up with the light beam. In short, there was no ether wind. It was as if they were riding in a car on a windless day, going 60 miles an hour and feeling no wind whatsoever when they stuck their arms out. Something strange was going on. Did the ether exist after all? What they had discovered was that you couldn't catch a light beam; in fact, chasing it would be useless. To see why, let's assume we set up a giant spotlight on Earth; when we turn it on, we know that a beam of light will travel out from it at 186,172 miles per second relative to the ether. At this speed the beam would take roughly eight minutes to get to the sun, or one-seventh of a second to go around the earth (if a system of mirrors were set up). Furthermore, if we examined the front of this beam we would see that there is nothing mysterious about it: the region directly in front of it would be dark, and the beam would, indeed, be traveling at 186,172 miles per second.

Let's assume, then, that we have a rocketship on a launch pad next to the spotlight. Assume that we blast off in the rocketship, at 186,000 miles per second, then a couple of seconds later somebody turns the spotlight on. It is, of course, pointed so that the light beam will pass directly beside the rocketship. What will we see? Within seconds we will see the light beam pass us and it will be traveling 186,172 miles per second *faster* than us. In other words, it will pass us as if we were sitting still. Even if we accelerated the rocketship up to 186,172 miles per second, the beam would still pass us at the same speed. Regardless of what we did, it would have no effect on the beam's speed.

It's no wonder the result stunned the scientific world when Michelson and Morley announced their results in July of 1887.[2] It was like having a car that traveled only at 75 miles an hour, but regardless of how fast you went in trying to catch it, it still went 75 miles an hour faster than you. In essence you could never catch it, even though it was going only 75 miles an hour. We know this doesn't happen with cars, but apparently it does with a beam of light.

ATTEMPTED EXPLANATIONS

Once the implications of the result had sunk in, everyone waited for an explanation. The most obvious one was that the earth somehow carried the ether along with it. But that would be like a car that carried the air around it with it. It didn't seem possible, and experiments soon showed that it was extremely unlikely.

Most scientists were in a quandary. There seemed to be no explanation. George Francis Fitzgerald, in Ireland, came up with what seemed to be one of the strangest explanations, but it worked. Born in 1851, Fitzgerald attended Trinity College in Dublin, graduating in 1871. After graduation he stayed on as a professor of natural philosophy. He postulated that the ether put a kind of "pressure" on everything that moved through it, causing it to shorten in the direction of travel. He even went as far as deriving a formula that gave the amount of contraction with velocity needed to explain the result.

What Fitzgerald was saying, in effect, was that a measuring rod such as a yardstick would shorten as it traveled thought the ether—the greater the velocity, the greater the shortening. But what was causing the shortening? The only explanation seemed to be the pressure on the rod from the ether, but this was hard to accept. Fitzgerald had, in effect, given us a formula that accounted for the Michelson-Morley experiment, but it was little more than a "fudge factor" with no explanation.

The shrinkage would not be very great at ordinary velocities. In fact we would never notice it. According to the formula a baseball shrinks slightly when it is thrown, and it is therefore no longer round. But the amount of shrinkage is negligible and would be impossible to measure. To see how great the shrinkage would be, let's consider the shortening of a yardstick (36 inches) at various speeds. At half the speed of light it would shrink to about 31 inches. At three-quarters the speed of light it would shrink to approximately 24 inches, and at 90 percent the speed of light it would be 15.6 inches long.

The best way to perform the above experiment would be to attach the yardstick to the side of a rocketship and have it pass overhead. Then, using some sort of optical instrument, you would have to measure the length of the yardstick. Of course it wouldn't be just the yardstick that shrunk; the rocketship itself would also shrink.

Hendrick Lorentz in Holland also came up with the formula for shrinkage just after Fitzgerald did.[3] He looked at the result very closely and tried to explain it. He considered the electric and magnetic fields in the rod, and the resulting forces in an attempt to show that they were somehow involved in the shrinkage. He also suggested that time intervals would undergo a similar shortening, or dilation. His theory was much more complete than Fitzgerald's, but in the end he wasn't able to explain things either.

STRANGE RESULTS

Looking more closely at Fitzgerald and Lorentz's formula we see something particularly strange. It predicts that a rod such as a measuring stick would get shorter and shorter as the velocity increases. But what happens when the velocity gets to the speed of light? According to the formula the rod disappears; in other words, it has zero length, and therefore no longer exists. Of course, if we got in a rocketship and traveled alongside it, it would appear normal. But to someone on Earth, both the rod and the rocketship would have disappeared.

It's no wonder scientists were baffled. There seemed to be something very strange about the speed of light. Everything disappeared when it went at the speed of light relative to us. Did this mean that the speed of light was the uppermost possible speed in the universe? We will look into this later.

The Michelson-Morley experiment was repeated many times over the next few years, but the results were always the same. Roy Kennedy and Edward Thorndike of the United States changed the length of the arms on the Michelson interferometer, but it didn't change the results.

The world waited for an explanation, and it came a few years later from an unexpected source: a patent examiner in Bern, Switzerland.

3 Special Relativity

O NLY A FEW YEARS IN THE HISTORY OF SCIENCE
stand out so much that they are referred to as
"miracle" years. The year 1905 was such a year. In
1905 Einstein published five monumental papers.
All were important, but among them was one that
literally changed the face of physics.[1]

At the beginning of 1905 Einstein wasn't concerned about changing the nature of physics as he
was getting his doctorate. He had finally landed a
job at the at the patent office in Bern and had
helped Besso get a job there also; his main concern
now was completing a dissertation or thesis, so he
could get his doctorate. He talked to Besso frequently about a thesis topic. He had been working
in several different areas, but some of them were
too controversial for a thesis. In the year 1900 the

German physicist Max Planck had introduced the idea of "quanta" or chunks of energy to explain a phenomenon related to heat. Einstein had become intrigued with it, but the idea had not yet been accepted and was very controversial, so Einstein was cautious about doing a thesis in the area. He had also done some work on the electrodynamics of moving bodies, but had already tried to use it as a thesis and it had been rejected. The only thing left, it seemed, was his work on atoms and molecules. To him it was unbelievable that several respected scientists still did not accept the reality of atoms. Ostwald, Mach, and others were sure the concept was merely a useful device for dealing with certain problems, but as far as they were concerned, atoms as such did not exist. Einstein was determined to show beyond a doubt that they did.

One day while he was having tea with Besso he began thinking about the problem. They had been talking about a possible thesis topic. As Einstein stirred his tea he realized the viscosity of the liquid increased as he put sugar in it. Was it possible that the viscosity was related to the size of the sugar molecule? That evening he looked into the problem more closely, and was able to derive a relationship for the size of the molecule in terms of viscosity and the diffusion rate (rate of spreading through a medium). During April of 1905 he wrote his thesis up and submitted it to his advisor, Dr. Kleiner, at the University of Zurich. It was seventeen pages long.[2]

A few days later he got it back with the note saying that it was too short. Einstein was annoyed. Going through the thesis he saw little wrong with it, but added one sentence and sent it back. Surprisingly, Kleiner now accepted it.

It was about this time that Einstein sent a letter to Habicht describing the work he was doing.[3] He mentioned four projects. "[The first] is on the radiation and energy of light, and is very revolutionary. . . . The second discusses the methods of determining the real dimensions of atoms. . . . The third proves . . . random motion [of small bodies in liquid] due to thermal motion of molecules. . . . The fourth is based on the concepts of electrodynamics of moving bodies, which employs a modification of the theory of space and time."

THE FOUNDATIONS OF SPECIAL RELATIVITY

With his thesis out of the way Einstein turned his attention to the last of the four papers. Although Einstein never referred to it as revolutionary in his letter to Habicht, it would indeed be revolutionary in that it would change the foundations of physics. Einstein talked to Besso about the problem, sometimes at work, but mostly when they walked to and from work, and in the evenings. He could see the problems clearly, but still wasn't sure how they could be overcome. The concept of time was his first hurdle. Newton had postulated an absolute time—a time that was the same throughout the universe. When he referred to an instant "now" there was a corresponding "now" throughout the universe, and they presumably all occurred at the same time. But to Einstein this didn't make sense. There was no problem relating "now" between two centers on Earth, but if you tried to relate them between a city of Earth and a star such a Vega, which was 26.5 light years away, there was a problem (see figure 3.1).[4] He knew that the image of Vega that we see is 26.5 years old—the time it takes for the light from Vega to reach us. And the only way we could communicate with Vega is via a radio signal, which travels with the speed of light.

We can't signal it with an infinite velocity, and therefore we have no idea what "now" is like on Vega. The only way we will ever know is to wait for 26.5 years. The concept of an instant "now" throughout the universe was therefore flawed. Time could not be absolute. To Einstein, time was merely an ordering of events. In our solar system, for example, everything is geared to the motions within it. Our day is the time for the earth to spin once on its axis; our year is the time for the earth to go around the sun. These times would make no sense for someone in a different planetary system.

Furthermore, two events in space that we see as simultaneous wouldn't necessarily be simultaneous for someone observing them from a different system. To see why, consider observers in each of two spaceships. We'll call them Pete and Mike. Assume that two explosions occur, one directly in front of Pete's spaceship and one

Fig. 3.1. Einstein looking at the star Vega.

directly behind (see figure 3.2). To Pete they appear as simultaneous. At the time of the explosion, however, Pete notices the spaceship with Mike in it pass him at a high speed. Mike also sees the two explosions. We ask: would Mike see the two events as simultaneous (as Pete did)? In other words, would they appear to Mike to occur at the same time? The answer is no. The explosion in front of the two spaceships would appear to Mike to occur slightly ahead of the one behind. The reason is that the light signal from the explosion in front would take a slightly shorter time to get to him, and the one behind a slightly longer time, because of his higher speed. What this means is that two events that are simultaneous in one system (it is convenient to refer to our spaceships as systems, or frames of reference) are not necessarily simultaneous in another

Fig. 3.2. Explosion occurring in front of and behind two spaceships. The spaceship on the left is traveling faster than the one on the right.

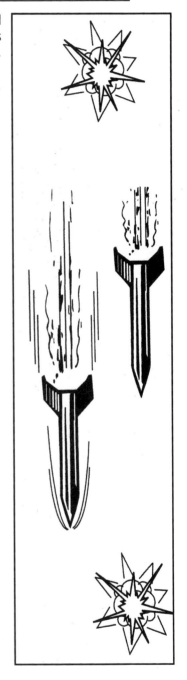

system. The realization of this was a significant breakthrough for Einstein. Once he had accepted it, he was able to formulate an entirely new theory of space and time.

It may surprise you that I haven't mentioned the Michelson-Morley experiment. Was Einstein not concerned with it? He had likely heard of it, even though in later life, he said he wasn't sure he had. (In his paper he wrote, "unsuccessful attempts to detect a motion of the earth relative to the light medium," which refers to it.[5]) But Einstein's major concern at the time wasn't the Michelson-Morley experiment. He was more concerned with a corresponding problem in the relationship between electric and magnetic fields and in Maxwell's theory of electromagnetism. As we saw earlier, an electrical charge is surrounded by an electric field. We can't see it, but it's easy to detect. If we bring another charge into the field, it feels this field; in other words, it is acted upon by a force. In the same way a magnet is surrounded by a magnetic field, and in 1820 the Danish scientist H. C. Oersted discovered a relationship between the two fields, namely, that a moving charge (which has an electric field associated with it) creates a magnetic field. Furthermore, in 1821, the English experimenter Michael Faraday showed that a moving magnet can produce a current in a wire.

Einstein was primarily concerned with

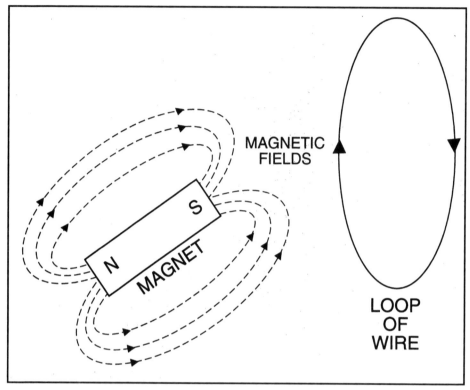

Fig. 3.3. Magnetic field of a magnet and a loop of wire.
When the magnet is moved a current flows.

an experiment you could easily perform using a wire loop and a magnet (see figure 3.3). If you hold the loop fixed and move the magnet through it, a current appears in the loop, and there was a clear, straightforward explanation for this current. On the other hand, if you held the magnet and moved the loop over it you also got a current; in fact the current was exactly the same. But the explanation of why the current appeared was completely different from the case where the magnet moved. This didn't make sense to Einstein. To him the only thing that was important was the *relative* motion between the magnet and the loop. But the relative motion wasn't important, according to Maxwellian theory. There were two explanations for the phenomena, and that's the way things were.

But Einstein had never been one to accept explanations without proving them for himself.

Like most breakthroughs, the understanding of this experiment came to Einstein in a flash of insight. He had been talking to Besso at length about simultaneity, absolute time, and absolute space. Then one morning, after thinking about them as he went to bed, he rose with all the pieces of the puzzle in place. As he greeted Besso that morning on his way to work he said, "The problem is solved."[6]

Einstein's insight was to see things clearly now for what they were. Newton was wrong. Even though it was almost blasphemy to say such a thing, Einstein was convinced he was right. There was no such thing as absolute motion or absolute time. The only absolute in the universe was the speed of light. It was independent of the speed of the source (as was shown by the Michelson-Morley experiment). Like it or not this was something that had to be accepted.

It might seem strange that a light beam passes you here on Earth with the same speed it passes a spaceship going at 185,000 miles per second. Somehow it doesn't seem possible, but if you think about it for a minute, constancy of the speed of light is important in the universe. Consider a binary system (two stars orbiting around their center of mass; see figure 3.4).[7] Suppose the earth is in the plane of revolution of the two stars so that we see one of the stars approaching us for a while (as the other recedes) then the other star approaches as the first recedes. Orbital speeds in a system such as this can be very high.

If speeds did add to that of light (as was assumed before the Michelson-Morley experiment was performed), the beam from one of the stars would travel to us at the speed of light plus the orbital speed of the system. The beam from the other star would travel to us at the speed of light minus the orbital speed. Then a little later things would reverse. In short, we would have many beams of light from this system traveling through space in our direction, each with a different speed, and as a result they would pass one another on their trip to us. Extrapolating this to the entire universe, where most objects are traveling at high speeds relative to us, we would have a hard time seeing it clearly. Einstein's postulate that the speed of light is a constant of the universe seems reasonable on

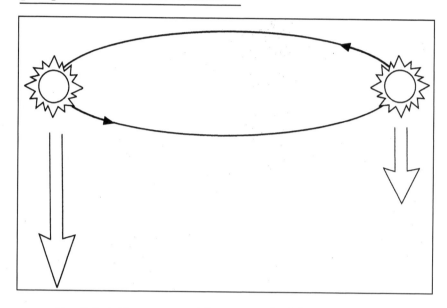

Fig. 3.4. A binary system. The star on the left is approaching. The one on the right is receding. If the star's speed added to the speed of light, the speed of the two beams would be different.

the basis of this, and of course it agrees with the Michelson-Morley experiment.

As a second postulate Einstein said that all motion in the universe is relative. Indeed, just as there is no such thing as a "now" throughout the universe, there is also no such thing as absolute motion. Only relative motion makes sense. But this is in conflict with the concept of the ether. According to Lorentz, all motion is measured relative to the ether. Since Einstein rejected absolute motion he also had to reject the ether. He referred to it as superfluous; in other words, it wasn't needed, and as far as he was concerned it didn't exist.

To understand the concept of relative motion more fully, let's assume that the universe has nothing in it except two spaceships, and that you are in one of them when suddenly you encounter the other, and pass it (see figure 3.5). You can easily measure the speed with which you passed it, but can you be sure you actually passed it? It's possible that you were standing still in space and the other spaceship passed you. On the other hand, both of the spaceships could be moving, each with part of the overall velocity. According to Einstein there is no way you can ever determine which of the above is true. All motion is relative to something, and without that "something," motion doesn't make sense.

Now let's consider time in relation to the two above spaceships. We saw earlier that events are not necessarily simultaneous for observers on the two spaceships. This implies that intervals of time are therefore also not necessarily the same, and therefore a clock on one of the spaceships would not run at the same rate as a clock on the other spaceship. But we would somehow have to relate the two clocks. Einstein realized he would need some way of relating time in one of the systems to that in the other system (spaceship). Furthermore, this relationship would also have to include space measurements. Newton and Galileo had given a relationship or "transformation" between systems such as this earlier, but Einstein knew that it could not be used. In the Galilean relationship we could assume one of the systems was at rest and add the speed of the second system to it. For example, if a car going 60 miles per hour passes one going 40 miles per hour, the faster one is obviously

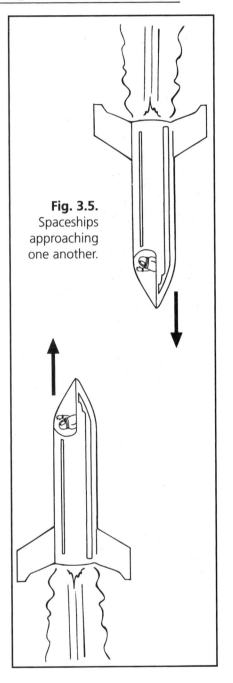

Fig. 3.5.
Spaceships approaching one another.

going 20 miles per hour faster than the slower one. Nothing could be simpler and more straightforward. Yet Einstein knew this wouldn't work in his theory; he needed a new relationship. Starting with his two postulates (constancy of the speed of light and relativity of motion) he was able to derive a relationship between two systems in motion. It was distinctly different from the Galilean relationship, and appeared to be more complicated.

Unknown to Einstein the same equations had been derived earlier by Lorentz in Holland. They were the equations we discussed earlier, the equations that gave a length contraction. The contraction in this case was assumed to be due to the "pressure" of the ether on our measuring rod as it moved through it. Lorentz had retained the concept of absolute space and time and he had retained the ether in deriving his relationships. So while the equations looked the same, Einstein's interpretation of them was completely different. Also, as we will see, Einstein went much further than Lorentz in explaining time, mass, addition of velocities, and energy.

CONSEQUENCES OF EINSTEIN'S POINT OF VIEW

One of the major predictions of Einstein's theory was a contraction in the length of rods at high speeds. But Lorentz and Fitzgerald predicted this, you say, so it was nothing new. True, it was predicted by these men but there was a lot that was new. The interpretation was entirely new—entirely different. Lorentz's contraction was a "physical" contraction due to the pressure of the ether. Furthermore, the velocity referred to in Lorentz's formula was a velocity relative to the ether, and therefore Lorentz's rod had an absolute "rest" length (its length while at rest in the ether); Einstein's did not.

To see one of the fundamental differences between the two interpretations, consider a spaceship flying above the earth with a yardstick attached to it that we can measure from Earth using a telescope (see figure 3.6). Lorentz's theory predicts that as it goes faster and faster we see the yardstick (and the spaceship) get

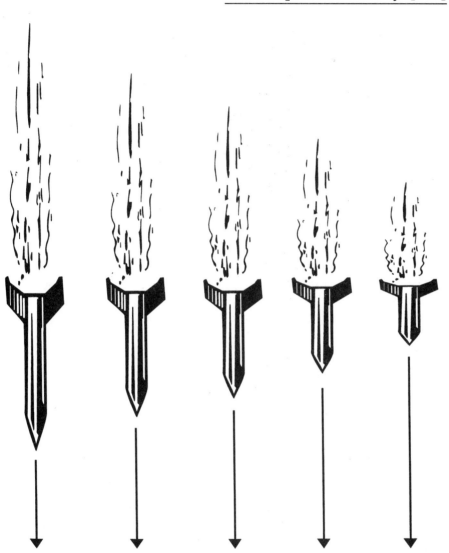

Fig. 3.6. Length contraction. Note that as the spaceship travels faster and faster (relative to us) it gets shorter and shorter in the direction of travel. The one to the right is traveling the fastest.

shorter and shorter in the direction of travel. But what does the observer in the spaceship see when he looks at a yardstick on Earth (again using a telescope)? He sees it get longer and longer, according to Lorentz. But according to Einstein the yardstick on Earth also gets shorter and shorter. In other words, both observers see it get shorter. Is this possible? To see that it isn't as strange as we might expect, consider an analogy. Assume we have a giant optical lens with an observer on one side and another observer on the other (see figure 3.7). We'll call them Pete and Mike again. Pete looks at Mike and sees him smaller than he really is, and Mike looks back at Pete and also sees him smaller than he really is. Both see each other as smaller and there doesn't appear to be a problem. Of course you might now ask: Is the contraction real or is it an optical illusion? It's important to point out that it is real. True, if you jumped in a spaceship and traveled alongside the first spaceship and measured the yardstick it would appear normal; it would be one yard long, and that is because the first spaceship no longer has a velocity relative to you. You are both traveling with the same velocity.

The reality of the contraction is easier to see when we consider time. Lorentz also postulated that there would be a time dilation, but he didn't look into it in detail. By time dilation, we mean that the clock on the spaceship will run at a different rate than one on Earth (it will run slower). Again we see a similar relationship to contraction. When the observer on Earth looks through his telescope at the clock on the spaceship he sees it slowed down, and when the observer on the spaceship looks through his telescope at the clock on Earth, he sees it running slow. Both observers see clocks running slow. This was postulated by Einstein, but not by Lorentz. The "reality" of this time dilation is easier to see than the length contraction because time intervals "remain" after the event, unlike length contractions which do not.

To see what I mean, consider our two observers again; Pete is in the spaceship above the earth and Mike is on the earth, and both are twenty-five years old. We know that if the spaceship is traveling at a high speed (close to the speed of light) Pete's clock will run much slower than Mike's. Assume, then, that Pete goes to a distant star and back and he takes one year. Using Einstein's for-

Fig. 3.7. Pete looking at Mike through an optical lens.

mula for time we can easily calculate how much time passed back on Earth while he was gone. If he traveled at, say, 99.9 percent the speed of light, considerable time would have passed—25 years. Therefore, when Pete gets back to Earth he will only be 26 years old, but Mike will be 50. This is something that can be seen, and it tells us that time dilation is no illusion. In the same way, length contraction is no illusion. Still, it isn't something "physical" that happened to the measuring rod—a physical contraction of the atoms. It's a result of the measuring process.

One thing I should point out here is that the velocity we are referring to has to be constant. In other words it cannot be accelerated motion; we will see later that this is a distinctly different case.

HEY! WHERE DID ALL
THE MASS COME FROM?

We've seen that there are length contractions and time dilations, but length and time are part of a triplet that is fundamental to physics: length, time, and mass. You probably associate mass with weight, and indeed when you step on a scale you are getting a measure of your mass. We refer to this as gravitational mass, but there is also another mass that is used in physics. It is known as inertial mass and is a measure of the resistance to motion. You don't use a scale to measure it. It is a measure of the force needed to set an object in motion. (We will see later that these two masses are equal.)

Getting back to our fundamental triplet, it seems obvious that if both length and time undergo a change with speed, so should mass, and indeed it does. Again consider a spaceship passing overhead. According to Einstein's theory, as its speed increases its mass increases. It becomes greater and greater, until at the speed of light (in theory) it becomes infinite. Since this is impossible it implies that the speed of light is unattainable. In fact, if we look at the formulas for length and time we see something just as dramatic happening at the speed of light. The length of a rod contracts to zero; in short, it shrinks to nothing at the speed of light. Furthermore, it's also easy to see that clocks stop at this speed. Again we have to be careful, however. This change is only evident to the observer on Earth; the observer in the spaceship notices nothing out of the ordinary, and indeed if the observer on Earth jumped in a spaceship and caught up with the first spaceship, he would not see the clock run slow.

We see, therefore, that as the spaceship (as viewed from Earth) gets shorter, it physically becomes more massive. In fact, when it is one-tenth its original length, it is ten times as massive. If it (in theory) becomes infinitely massive and at the same time disappears at the speed of light we obviously have a serious problem. It means that we can't travel at the speed of light, or more generally, no material object can travel at the speed of light. We can see this more clearly if we think of the amount of energy it takes to accelerate a spaceship to high speeds. If the mass of the spaceship is infinite at the speed of light, it obviously takes an infinite amount of energy to attain it, and there isn't an infinite amount of energy in the universe.

Of the three dramatic changes with speed, the change in mass is the easiest to observe experimentally. In fact, scientists observe it every day around the world in high energy accelerator labs. When a particle such as an electron or a proton is accelerated in an accelerator, its mass increases, and as its mass increases it takes more energy to accelerate it further. That's why we can take a particle such as a proton up to 90 percent the speed of light with a relatively small accelerator, but to get it to .9999 the speed of light it takes one that is many miles across.

We are also now able to experimentally detect time dilation. All particles except the electron (and possibly the proton) decay in a short period of time. We can, in fact, generate particles in the lab and measure their decay time. If these same particles are traveling at high speeds relative to us, say, striking the earth as high speed cosmic rays, their lifetime is extended according to Einstein's formula. In fact, it is an excellent verification of the formula.

If there are problems at the speed of light, it's natural to ask what happens beyond the speed of light. Looking again at Einstein's formulas we see that something very strange occurs. They give what mathematicians call an "imaginary" result. Imaginary numbers are not new; mathematicians deal with them every day, but what they mean physically is uncertain. A number of physicists, including Gerald Feinberg of Columbia University, have speculated, however, that there is a whole world beyond ours, populated only by particles that move at speeds greater than the speed of light. Technically, Einstein's theory says that material particles cannot travel *at* the speed of light, but it says nothing about speeds beyond that of light. Feinberg refers to these particles as tachyons, and as you might expect, searches have been made for them, but no one has found them. Most physicists do not take them seriously.

ADDING VELOCITIES: TWO AND TWO EQUALS FOUR ... DOESN'T IT?

We stated that the velocity of light was the greatest possible velocity for material objects in the universe. Nothing can travel as fast, or faster. But if you think about it for a minute you can easily come up with a velocity that is greater than that of light. Picture two spaceships approaching one another; both are traveling at 80 percent (or .8) the speed of light. They are obviously approaching one another at 1.6 times the speed of light. Is this possible? Einstein showed that it isn't. In addition to his other formulas he also gave one for the addition of velocities. You might think this was hardly needed; after all, if one car is going 60 miles per hour and another 40 miles per hour in the same direction, the first is going 20 miles per hour faster, and no one can argue with that. Einstein showed, however, that when you deal with velocities near that of light this is no longer true. Things don't add up in the usual way. In fact, according to his formula they don't add in the usual way even at low velocities, but like length contraction and time dilation, the effects are so small at low velocities we never notice them. But at velocities close to that of light this is not the case. Einstein's formula shows, in fact, that you can never get a velocity greater than that of light by adding two other velocities. Furthermore, the formula shows that when you add a velocity to that of light it just

Fig. 3.8. "What do you mean . . . 4? That's not what Einstein said."

gives the velocity of light back. So again we have another proof that the velocity of light is the limiting velocity in the universe.

Now let's go back to the problem that started it all: the Michelson-Morley result. In that experiment they tried to add the orbital velocity of Earth to the velocity of light and got a null result. In other words, they got the velocity of light and nothing more. It was as if the velocity of the earth had no effect. But this is what our addition of velocities formula tells us we should get, so Einstein's theory adds another feather to its cap; it explains the Michelson-Morley experiment.

Interestingly, there was another experiment that was performed by Armand Fizeau even before the Michelson-Morley experiment. In it light was projected along a tube of running water. As in the case of the Michelson-Morley experiment, Fizeau showed that the speed of the water had no effect on the speed of the light beam, and Einstein's theory also explains it.

MAXWELL'S THEORY

Earlier we saw that Einstein was less concerned with the result of the Michelson-Morley experiment than he was with the problems in Maxwell's electromagnetic theory. In the problem of the magnet and the loop of wire, the interpretation was different, depending on whether the magnet or the loop of wire was moved. To Einstein the only difference was one of relative motion, and indeed his theory showed that this was the case. In fact, according to his theory, electric and magnetic fields are related; they appear different to us only as a result of their motion, and again the speed of light plays an important role. If we start with an electric field at rest relative to us, it will, of course, appear as a "pure" electric field. In other words, there is no magnetic component. If we move it relative to us, however, it will be both an electric and a magnetic field. Furthermore, the faster we move it, the greater the magnetic component will become, until finally at the speed of light it will be a "pure" magnetic field, with no electric component. Therefore, whether or not a field is electric or magnetic depends on our motion relative to it. Einstein also showed that the mathematical form of Maxwell's theory was the same, regardless of the motion. In Lorentz's theory this was not the case. If you were at rest relative to the ether, the equations had a simple and beautiful form, but if you had a velocity relative to the ether, they changed. To Einstein they became ugly, and he disliked the change. With his new theory they were always the same.

ENERGY

Einstein wrote up his new theory and submitted it to *Annalen der Physik* in June 1905.[8] He was worried it might be rejected because it was so revolutionary. After all, if it was correct, the foundations of Newton's theory would crumble. But it was refereed, checked for accuracy, and published in September, 1905. It was all of twenty-nine pages long.

With such a powerful paper under his belt you might think

Einstein would rest a while, but his mind was still whirling with ideas. So far he had dealt with length, time, and mass, but what about energy? Motion was a form of energy, and mass was related to energy. When a mass was in motion it had what is called kinetic energy. Einstein examined energy carefully in light of his new theory and found something astounding: energy has mass. This was important, for example, in relation to radioactive decay. Indeed, it explained the phenomenon. Of even greater importance, however, Einstein later showed that this could be reversed: mass had an energy associated with it. In short, mass and energy were two forms of the same thing. In theory, but not necessarily in practice, a given mass could therefore be converted to energy. The equation relating the two phenomena can be written out as energy equals mass times the speed of light squared (the speed of light times itself).[9] Since the speed of light is a large number, the square of it is huge. There is therefore a tremendous amount of energy associated with even a small amount of mass. Most people know that this is the basis of the atomic and hydrogen bombs. Indeed, it cleared up a mystery that had been around for years: the energy of the sun and stars. Where did it come from? According to Einstein's theory it came from mass that was continually being converted to energy in stars. It was this energy that kept them going. With so much energy given out by a small amount of mass, we don't notice any mass loss in the stars.

WAITING FOR A REACTION

After submitting his paper on the electrodynamics of moving bodies, what is now referred to as special relativity, Einstein waited for a reaction. He knew it would be controversial, and he was ready for the criticism. But over the next few months he heard nothing. Nothing was said about it in the next few issues of *Allanen der Physik*. But of course information like this takes time to assimilate and digest. According to Einstein's sister's (Maja) memoirs, he was bitter at the silence, but there is no indication from him of his bitterness.[10] After a few months he got a letter from Max Planck, asking for several clarifications of the theory.[11] Einstein was delighted; it meant the paper was being read. He kept up a correspondence with Planck over the next few years. Planck was already actively organizing seminars in Berlin on the new theory. Planck's assistant, Max von Laue, attended one of these seminars and was soon intrigued with the new theory. He traveled to Bern to meet Einstein, expecting him to be at the University of Bern. He was quite surprised when he found him at the patent office.

As news of the new theory spread, others began requesting reprints, and as expected criticism soon followed. Many well-known scientists, however, strongly supported the theory. In the early 1920s the Nobel committee in Sweden decided to award Einstein the Nobel Prize, but even as late as 1920 special relativity was still controversial and they had to award it to him for one of his other contributions—his work on quantum theory. With advances in technology during the 1920s, however, Einstein's theory could be checked much more easily and soon it was accepted universally.

4

Four-Dimensional Space-Time and Time Travel

H OW COULD EINSTEIN'S SPECIAL THEORY OF relativity be checked? One of the easiest predictions to test was the increase in mass with speed. But Einstein's wasn't the only theory that made such a prediction. Max Abraham of Göttingen had also presented a theory of the electron that predicted a change in mass. And another theory had been put forward by Heinrich Bucherer. Walter Kaufmann of the Bonn decided to see which of these theories appeared to be most consistent with observation. In January 1906 he published his results. According to them, Abraham and Bucherer's theories fit observations fairly well, but Einstein's theory didn't.[1]

What was Einstein's reaction? He wasn't worried; he knew it was only a preliminary experiment, and he was confident his theory would eventually be shown to be correct. And indeed, within a year he was proven right. Bucherer redid the experiment with a much more accurate apparatus and showed Einstein's prediction was correct.

During this time Einstein was still working at the patent office. As his work became better and better known, however, he started to become impatient for a better position—a position at a university. Although there was some interest, nothing was offered and he began to get frustrated.

Even more attention was directed to relativity theory after 1908 when Einstein's old teacher, Hermann Minkowski gave it a new formulation. Minkowski was no doubt surprised to see such an important paper on space and time published by his former student, a student he didn't particularly like. Einstein took a large number of math courses from Minkowski, but he skipped classes so often Minkowski rarely saw him. Einstein didn't like the way Minkowski lectured; to him, Minkowski's mathematics was abstract and seemed to have little application to physics. Furthermore, Minkowski never used examples, and according to Einstein he seemed to care little if his students understood him.[2]

In his fourth year Einstein was pleased when Minkowski offered a class titled "Applied Mathematical Physics." He quickly signed up for it, but was again disappointed; there was little mention of physics in it. Not until the very last lecture of the year did Minkowski finally give a lecture that appealed to Einstein. It was on capillary action (the rise of liquids through small diameter tubes). "That's the first lecture on theoretical physics I've had at the Poly," he said to a friend.[3]

Despite Minkowski's apathy towards Einstein as a student, his modification of Einstein's theory was a monumental contribution. Born in Russia in 1864 to German parents, Minkowski obtained his doctorate at the University of Königsberg in 1885. A few years later he began teaching at the Polytechnic in Zurich, where Einstein was later a student. Two years after Einstein graduated, Minkowski was invited to Göttingen where he worked with the world famous mathematician David Hilbert. Both men were convinced that theoretical physics was too difficult for the average physicist. They therefore spent considerable time studying basic physical concepts such as space and time. As expected, Einstein's paper eventually came to their attention, and they realized it was a particularly important paper. According to Minkowski's assistant Max Born, Minkowski couldn't believe that Einstein had written it. He didn't think he was capable of it.[4] Nevertheless, he was impressed. After studying the paper carefully, he saw that it could be formulated much more elegantly by assuming time was a "fourth dimension" of a "continuum" he referred to as space-time. A continuum is something where the points are so close together they are continuous; in essence there are no breaks in it. Ordinarily space is a continuum.

SPACE-TIME DIAGRAMS

In order to understand Minkowski's contribution, it is best to start at the beginning, in other words with Newton's laws and simple space-time diagrams. It's important to remember that space and time were separate in Newton's theory; space had three dimensions and time one dimension. Also, distances in space were the same for all observers, regardless of their motion, and time was the same for all observers. Einstein showed that this was not true. For two observers in motion with respect to one another, they would be different.

To see why, let's use a diagram. We'll begin with what is called an "event." An event is anything specified by a place and a time. For example, if you meet someone at 4th and Main Street at 5:00 P.M., it is an event. You have specified the place and the time. On Earth we need only two dimensions (4th and Main Street, which are perpendicular to one another) to specify a place because we are dealing with the surface of the earth. In reality there is a third dimension, namely, height. Space, then, has three dimensions: right-left, up-down, and forward-backward. Time, on the other hand, moves in only one direction, namely, the forward direction, so it is one-dimensional.

We can make plots of space and time by assuming time is the vertical and space the horizontal. Space, as we just mentioned, is three-dimensional, but for convenience we represent all three dimensions by one.

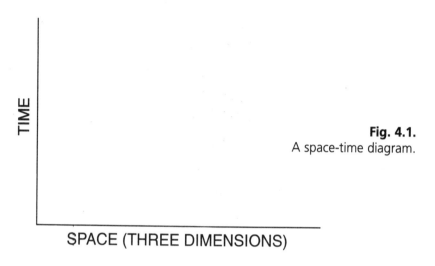

Fig. 4.1.
A space-time diagram.

An event in this diagram is a point. It specifies both a time and a distance (space). More generally we can use such diagrams to plot a trip. Suppose we want to make an automobile trip from Seattle to Denver, with a few stopovers. If we stop over at Boise and Salt Lake City our trip will look like this:

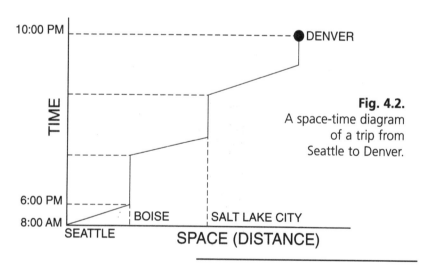

Fig. 4.2.
A space-time diagram
of a trip from
Seattle to Denver.

We start out at Seattle at 8:00 A.M., arriving at Boise at 6:00 P.M. where we stay over. Our stay over is represented by the vertical line (time passes but no distance is traveled). The next morning we leave for Salt Lake City. Note that our speed is represented by the slope of the line. Finally we arrive in Denver the next day.

If we took a similar trip in three dimensions, an airplane trip, for example, we would need to specify the altitude at all times. As a simpler example, consider the three-dimensional trip of a fly in a room. Let's say it starts out in the comer of the room and flies to

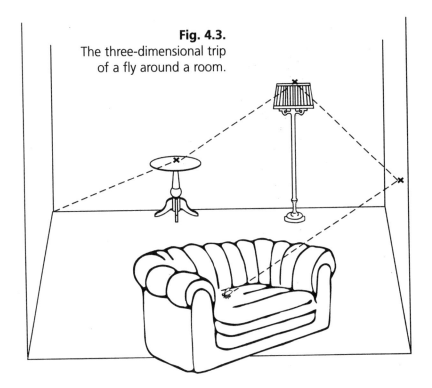

Fig. 4.3.
The three-dimensional trip
of a fly around a room.

the top of a stand where it stays for a few minutes, then it flies to a lamp, where it stops again. Then it flies to a wall and finally to the couch where it gets swatted. There's a problem, though. So far we don't have time in this diagram. For this we would have to make another diagram (see figure 4.4).

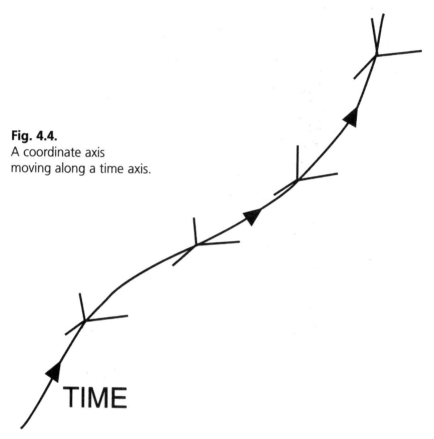

Fig. 4.4.
A coordinate axis
moving along a time axis.

TIME

The positions of our room and the fly are shown along the continuum of time. This is obviously inconvenient and we don't usually do it.

As we saw earlier, Minkowski brought space and time together into a four-dimensional space-time continuum. How could we represent events in it? It's hard if not impossible to imagine four-dimensional space. Even Einstein couldn't do it, so I wouldn't worry about trying to visualize it. Still, we can represent four-dimensional space-time in a simplified diagram. We assume all three space dimensions are on the horizontal axis, and time is on the vertical one, as before.

Minkowski emphasized that we should not consider space and

time as separate entities. They should be considered together. His famous announcement was made in Cologne, Germany, in September 1908. The opening words of his talk are now famous: "Henceforth space by itself and time by itself, are doomed to fade away to mere shadows, and only a kind of union of the two will preserve an independent reality."[5]

One of the most important concepts in Minkowski's interpretation was a new "invariant." An invariant is something that remains the same at all times; in other words, it remains forever unchanged, like the speed of light. Minkowski showed that there is a space-time invariant in his four-dimensional space-time. Previously space and time had independently been invariants, but this was no longer true in Einstein's theory. To see how we get this new invariant consider a wall. Assume we mark a point along the vertical, then another along the horizontal as shown and join the two points. We then have a "space distance" (see figure 4.5).

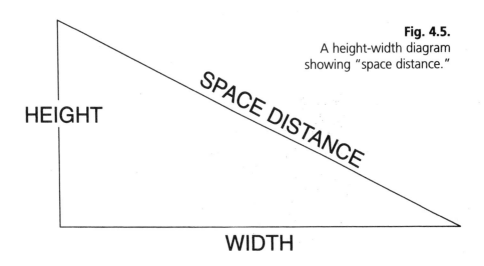

Fig. 4.5.
A height-width diagram
showing "space distance."

HEIGHT

SPACE DISTANCE

WIDTH

In the same way if we mark a time along the vertical axis in a space-time diagram we have a "space-time distance" (see figure 4.6).

You likely recognize this in relation to a famous theorem that

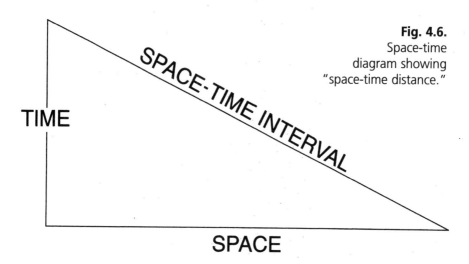

Fig. 4.6.
Space-time
diagram showing
"space-time distance."

you learned in school: the Pythagorian theorem. It tells us that if we have a right angle triangle such as the one above and know the length of the vertical and horizontal sides, we can easily calculate the length of the slanted side (what we call the hypotenuse). The relationship is: (side 1)^2 + (side 2)^2 = (hypotenuse)^2. See figure 4.7.

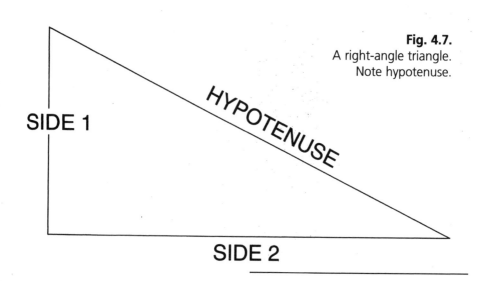

Fig. 4.7.
A right-angle triangle.
Note hypotenuse.

If we use this formula in our space-time diagram we can calculate the "space-time interval" shown. In this case $(space)^2 + (time)^2 = (space\text{-}time\ interval)^2$. There is, however, a problem. To make sense, space and time have to have the same dimensions. We can, however, change time into a space distance by multiplying it by a velocity. Remember, the distance you travel is speed multiplied by time. But what speed do we use? Minkowski chose the speed of light, which was an obvious selection, considering it is a constant of the universe. Our relationship is then: $(space\text{-}time\ interval)^2 = (space\ interval)^2 + (speed\ of\ light \times time)^2$. In practice, because time is quite different from space, we actually have to use a minus sign.

This is what we call a "proper" interval in four-dimensional space-time, but what is particularly important is that it is the same for all observers. To see the significance of this, let's go back to our two spaceships (figure 3.2). Pete, the observer in one of them saw two explosions, one ahead and one behind him occur simultaneously, but Mike, the observer in the other spaceship that was passing Pete, saw them occur at different times (and at different distances). The "space-time interval" between the observer and the explosion calculated as above, however, would be the same for both Pete and Mike.

How is this possible? It is easy to show using a diagram. In figure 4.8, we see that the two observers have different spaces and times, but they have a common space-time.

Banish Hoffman, in his book *Creator and Rebel*, uses a nice

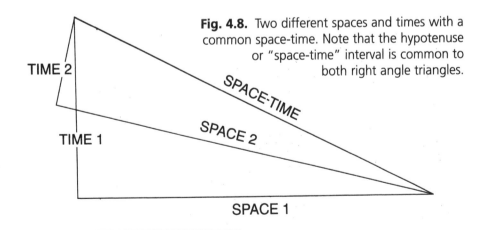

Fig. 4.8. Two different spaces and times with a common space-time. Note that the hypotenuse or "space-time" interval is common to both right angle triangles.

TIME 2

TIME 1

SPACE-TIME

SPACE 2

SPACE 1

analogy.[6] He says space-time is just like a piece of cheese that is being sliced differently by different observers. They both have the same space-time, but they slice space and time up differently.

We now have two invariants: the speed of light and the space-time interval between two events.

Let's turn now to how physicists represent space-time events. Again time is vertical and the three dimensions of space are horizontal. This time, however, we scale the graph so that a line at 45 degrees represents the speed of light (see figure 4.9). The origin of the graph represents "now," with anything above it being the future, and anything below the past. Any point within the upper region enclosed by the 45-degree line is an event, as is any point in the similar region below. The shaded, or dotted, region beyond the 45-degree line is called "elsewhere" because it is a forbidden region. It would take a speed greater than that of light to get there, and since the speed of light is the uppermost possible speed, we cannot get to these points. A sequence would be a line, and in our diagram we refer to it as a "world line." It is the "line" of the events that occurs in a short period of our life. If we take a trip in our spacetime, we plot it as a world line. Note that there are three types of trips.

The first (represented by A) is called a spacelike trip, and since it is at an angle less than 45 degrees, it is an impossible trip. The

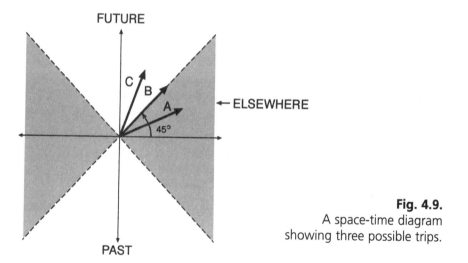

Fig. 4.9.
A space-time diagram
showing three possible trips.

second trip (represented by B) is called a lightlike trip, and is possible only for light. The third (C) is a timelike trip and for us the only possible trip. Therefore, our world line must be timelike in all parts.

We would like to generalize this so that the three dimensions of space look more like three dimensions. In practice we cannot generalize to three dimensions, but we can go to two by spinning the space dimension around the time axis. In this case we get two cones (see figure 4.10).

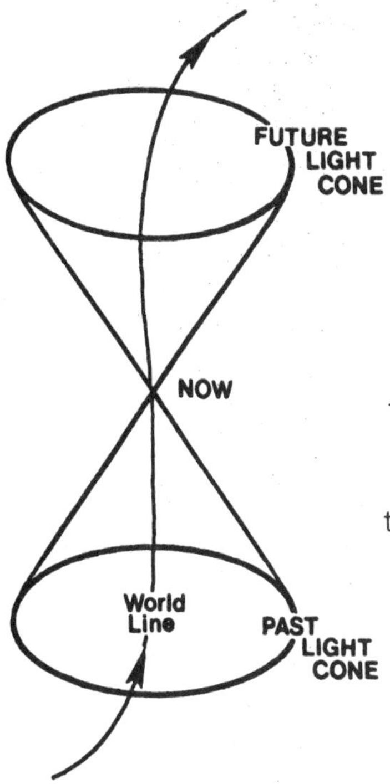

Fig. 4.10. A three-dimensional representation of figure 4.9. The line from the bottom cone through NOW to the top cone is called the world line.

TIME TRAVEL

Time travel is fascinating to contemplate. What would it be like to visit the future or the past? Is it, in fact, possible? As we saw earlier, we can easily look back in time; we merely have to look out into the universe. If a star is 20 light years away, we see it as it was 20 years ago. But we're of course looking for more than this, and we know that we have some control over time according to special relativity. The clock of an observer moving relative to us runs at a different rate than a clock that is stationary. Consider twins; we'll call them Bob and Fred. They are both 25. Suppose Bob jumps in a spaceship and travels to a nearby star and back, and it takes him one year. How much time will have passed back on Earth? This depends, of course, on his speed, so we'll consider several speeds, starting with 90 percent the speed of light. At this rate 2.3 years will have passed back on Earth while he was gone. Bob will therefore be 26 when he returns, but Fred will be 27.3. Not a large difference.

Stepping the speed up to 99 percent the speed of light, we find 7.14 years will pass, so Bob will be 26 and Fred 32.14. If we go to 99.9 percent the speed of light, 22.7 years will pass and Fred will be 47.2. Quite a difference this time. We have to remember, though, that Bob can't travel at the speed of light relative to Fred, but in theory he can get as close to it as he wants. At 99.99 percent the speed of light Fred will be 96.4 when Bob returns, and so on. It would, of course, be extremely difficult to get a spaceship to travel at 99.99 percent the speed of light, but as we will see there is an even greater problem that is frequently referred to as the twin paradox.

Assume Fred has just gone to a star and back at 99.9 percent the speed of light. He jumps out of his spaceship and says to Bob, "Ha! Ha! I'm only 26 and you're 47." Bob says, "No, you've got it all wrong. Einstein said all motion in the universe is relative. You and your spaceship actually stood still, and it was the earth that moved out to space and back at high speed, so I'm 26 and you're 47."

Fred looked at him in surprise. He was right; it was possible.

The obvious question is: Who is the younger when they get back together again? This is referred to as the twin paradox. The reason we have what seems to be a paradox is that we are doing some fudging. Einstein's theory of special relativity deals only with straight-line, uniform motion. It says nothing about accelerated motion, and if you look at our twins you find Bob, who is in the spaceship, has to accelerate to get up to speed, then has to decelerate and accelerate again when he gets to the star so he can turn around. We are therefore applying the theory to a situation that it doesn't cover. Who, then, is the youngest? The problem was cleared up in 1915 when Einstein extended his special theory of relativity to accelerated motion. According to this extension, which we will see is referred to as general relativity, the twin that underwent the acceleration will be the younger of the two. In short, if they are to move apart, one must undergo acceleration and he will be the youngest.

From this we see that we can, in theory, visit the future. Bob is

Fig. 4.11. "Ha! Ha! I'm 25 and you're 90."

still 26 when he gets back to Earth and everyone else is almost 23 years older. What about visiting the past? Could we go back and visit out great grandfather when he was a young man? There doesn't appear to be a way according to special relativity, but we will see later that it *may* be possible. We have to be careful, though. There's a principle called causality that we can't violate. According to this principle the cause of any event has to come before the event itself. In other words, cause comes before effect. If we throw a baseball and somebody hits it, the ball has to be thrown before the batter can hit it. Similarly if you pull the trigger of a gun a bullet fires, and again you have to pull the trigger before the bullet comes out of the barrel. It would be a strange situation if the bullet went out of the barrel first, then you pulled the trigger. Such a situation obviously violates causality.

Causality is particularly important in relation to time travel.

One of the most famous of causality violations is called the Grandfather Paradox. In this case you find out, say, that your grandfather was an evil man and did a lot of harm, so you decide to travel back in time and kill him, saving the world from the harm that he perpetrated. Of course, if you killed him when he was young, your parents would never have been born, and therefore you would never have been born, and if you were never born you couldn't have gone back and killed him.

Is there any way around the this paradox? As you can likely guess, a great deal of speculation surrounds it. One possibility for getting around it is that everything that could possibly happen in such a case actually does happen. In other words, the grandfather is both killed and not killed. That sounds crazy, but some scientists take it seriously. It is possible in what are called parallel worlds. At a point just before the killing (or nonkilling) there is a branching into several worlds, all of which pass through time parallel to our own world, but in different dimensions. In each of these worlds there is a different outcome to a significant event. We will later see that there are other equally strange possibilities.

TIME TRAVEL AND SCIENCE FICTION

One of the first, and still one of the most famous, time travel science fiction tales is H. G. Wells's *The Time Machine*, which was written in 1895.[7] This, of course, predates Einstein's special theory of relativity, so it couldn't have had any influence on Wells. Although it wasn't the first time travel tale it was the first to be based, at least slightly, on science. Well's traveler builds a machine that is able to take him to the past and the future, and return him safely to the present. The hero visits the past, but no mention is made of the Grandfather paradox, perhaps because Wells never thought of it, or knew about it. Anyway, the traveler never tampered with the past or tried to change it, so there was no problem.

Another well-known time traveler tale is Gordon Dickson's "Time Storm." In this story, as the title suggests, there is a violent storm in the time continuum that brings about a slippage of time on Earth, and needless to say, this slippage has some strange consequences.

A completely different point of view is presented in Paddy Chayefsky's *Altered States*. In this book our evolutionary past is presumably encoded in our genetic makeup, in other words, our DNA. Through the use of drugs and sensory techniques the hero of this story is able to bring his past to the present. To do this, however, he is forced to change in form to an organic soup.

Richard Matheson's novel *Bid Time Return* was made into a movie in 1980. The movie was called *Somewhere in Time* and it starred Christopher Reeve and Jane Seymour. In this case the hero is intrigued by a painting, and travels to the past (in a mysterious way) to meet the woman in the painting.

The idea of parallel worlds is presented in Clifford Semak's *Ring Around the Sun*. In this case there are many Earths in parallel universes, each presumably in a different dimension. The trick is to cross through these dimensions and enter these other worlds. The idea of parallel universes was also used in David Duncan's *Occum's Razor*.[8]

One of the most intriguing time tales was written by the cosmologist Fred Hoyle.[9] Since he is a scientist, the story contains a considerable amount of factual science. His speculation, however, is on par with most others. He begins by throwing out the concept of time as a smooth flowing, never-ending river. He replaces it with a continuum in which all time is equal, and the "present" plays no important role in it. According to Hoyle, it is consciousness that gives us a sense of the passage of time, and therefore it is an illusion.

I have only touched on the numerous books and movies in which time travel is used, but we'll come back to the concept later.

TRIP TO A STAR

Besides making the idea of time travel possible, Einstein's theory had a significant effect on our ideas of what a trip to a star would be like. It's unlikely that much attention was paid to what it would be like to travel to another star before Einstein's theory, but as we will see the theory had several effects.

The first effect is called aberration. It was discovered by James Bradley about 1725. The best way to understand it is to assume you are standing in a rain storm holding an umbrella. The rain, of course, falls straight down, and if you are standing you would hold the urnbrella directly over your head. If you start moving, however, the rain appears to be coming down at a slight angle and to remain dry you must tip the umbrella slightly in the direction you are going. In the same way, because we are on a moving platform (the earth) as we look out at the stars, we must tilt our telescopes slightly to allow the starlight to pass through it. In short, the star appears slightly ahead of where it really is. The angle is extremely small because the earth is not traveling very fast compared to light, nevertheless it is finite.

In the same way, if we were traveling at high speed in a spaceship, stars would appear to be out of position. If we were, for example, traveling in the direction of the star Vega, the stars that appear close to it would crowd even closer to it. As we went faster and faster more and more stars would crowd into our field of view. In the same way, if we could look out the back of our spaceship we would see fewer and fewer stars.

Interestingly, we would also see the color of these stars change due to what is called the Doppler effect. You are likely most familiar with this effect in relation to the horn on a car, or whistle of a train. As a car approaches blowing its horn you hear a distinct rise in pitch (frequency) as it approaches and a lowering of pitch after it passes and recedes. This is because of a crowding together of the sound waves as it approaches and a separation of them as it recedes.

The Doppler effect also occurs with light. The crowding together of the waves of an approaching light source is seen as a shift toward the blue end of the color spectrum. The increased separation is seen as a shift toward the red end, and is called a redshift. This means that stars in the direction that a spaceship is traveling will be blueshifted and those in the opposite direction will be redshifted.

In addition to this, the atoms that emit the light are, in effect, little vibrators, or clocks. And we know that clocks slow down as they move at high speeds relative to us. Therefore, as we approach the star, the vibrations will slow down and there will be a blueshift as in the case of the Doppler effect. Similarly, the stars behind the spaceship will be redshifted. This effect is more significant than the Doppler effect as we approach the speed of light. As the spaceship travels closer and closer to the speed of light the fraction of the sky in which a blueshift occurs becomes smaller, and the fraction over which redshifts occur become larger. So all in all we would have some strange effects which would seriously affect our view.

A PROFESSOR AT LAST

Needless to say, Einstein's special theory of relativity had a dramatic effect on physics. It took a number of years, however, for its importance to be fully realized. In the meantime Einstein lingered on at the patent office in Bern. In all, he remained there for four years after his theory was published. Many people thought it was crazy that he hadn't been offered a professorship, but the problem was that there were few positions available, particularly for theoretical physicists. Professor Kleiner at the University of Zurich had his eye on Einstein and had been hoping to hire somebody to help lighten his load. His first choice, however, was a former assistant,

---

What Einstein Does in His Spare Time

Einstein quote from 1918:

"When I have no special problem to occupy my mind, I love to reconstruct proofs of mathematical and physical theories that have long been known to me. There is no goal in this, merely an opportunity to indulge in the pleasant occupation of thinking."[10]

Friedrich Adler. But Adler only dabbled in physics, and was more interested in philosophy and politics. As Kleiner finally began to realize that Adler might not make a good addition to the physics department, his attention turned to Einstein. But there was a problem: Kleiner had seen Einstein lecture and wasn't impressed with his teaching abilities. Nevertheless, his research was making him well-known, and Kleiner knew he would be a valuable addition to the faculty at Zurich. Einstein was therefore invited to Zurich to give a lecture. He did a much better job than he had done previously and although Kleiner still had reservations he recommended him. When the vote was taken Einstein got ten votes in favor of him, with one abstention. He was invited to report for work the following fall in 1909.

5 General Relativity

INSTEIN WAS PLEASED WITH HIS special theory of relativity, but he knew it wasn't complete. It accounted for straight-line uniform motion, but it did not account for non-uniform or accelerated motion. And as everybody knows, accelerated motion is common on Earth and throughout the universe. Jump into your car and you have to accelerate to get it up to speed. In fact, even when you're going around a corner at what seems to be a uniform speed, you're actually accelerating. Einstein knew he had to extend his theory so it would cover non-uniform motion.

The theory was also incomplete in another sense. It said nothing about gravity. It wasn't clear at first that gravity would be included in a generalization of the theory, but a sudden insight Einstein had in 1907 led him to believe it had to be. He was writing a review article for the *Yearbook of Radiation and Electronics* on his special theory of relativity when he realized that he would have to say something about the incompleteness of the theory.[1] While contemplating what he should write, a thought struck him: if someone fell off a high building they would feel no gravity on their trip to the ground. They would, of course, be accelerating with the usual acceleration of 32 feet per second per second. Did this mean that there was a relationship between acceleration and gravity? It appeared so. Thinking about it further he concluded that if he could extend his theory to acceleration it would also be a gravitational theory.

Einstein had, for years, disliked Newton's theory of gravity. According to Newton, two objects attracted one another across space with a strange action-at-a-distance force. Move one object and the other moved. It was too mysterious, too artificial, for Einstein. He was sure there was a better way. The link, Einstein was now sure, was acceleration.

Still, acceleration seemed to be absolute. If you accelerated, you always felt a backward push into your seat. Even if you closed your eyes you knew you were accelerating. Somehow it didn't seem possible that accelerated motion was relative.

Another way of looking at this is to think of somebody in an elevator in a high building. If the person stands in the elevator when it isn't moving, everything will be normal. If he steps on a scale, he will weigh what he normally weighs. If he jumps, he will rise from the floor in the usual way, and if he drops a ball, it will fall to the floor in the usual way. But what happens if the cable holding the elevator suddenly breaks and the elevator begins free falling? The man will now be weightless. If he gives himself a slight shove, he will float around the elevator. We're used to seeing this type of thing nowadays with TV pictures of astronauts floating around in spaceships, but Einstein was visualizing something new and strange.

The falling elevator would be like an elevator out in space, far from any gravitational influence. But what if we now attach a cable to its top and begin pulling it upward with an acceleration of 32

Fig. 5.1. Einstein in an elevator out in space being accelerated upward.

Fig. 5.2. Einstein in an elevator sitting on Earth.

feet per second per second—the acceleration of gravity on Earth? The elevator floor would then be pushed against your feet and you would feel like you were on Earth. It would seem as if you are in an artificial gravitational field. Einstein thought deeply about the problem, finally coming to the conclusion that it wasn't an artificial field that was being created by the accelerating elevator. It was a "true" gravitational field. In other words, gravity and acceleration were intricately related. Einstein later referred to this as the "Happiest thought of his life."[2]

INERTIA AND GRAVITY

To understand Einstein's breakthrough better, let's turn back three centuries to the time of Newton. One of Newton's major accomplishments was the formulation of a law of inertia. According to this law, any object in the universe continues in a straight line in uniform motion unless acted upon by a force. If the object is subjected to a force its inertia will be overcome and it will start moving in a different direction. As I mentioned earlier, you feel a force pushing you back against the seat of a car when the car is acceler-

Fig. 5.3. Astronaut in space pushing bowling balls of various mass.

ated. This is the force of inertia. Things at rest don't want to be moved and resist any attempt to move them. They fight back with inertia. That's why it takes quite a push to move a stalled car. It is easy to calculate the magnitude of the force needed for any acceleration from another of Newton's laws, assuming you know the mass of the object you are pushing. Or, turning things around, you can calculate the mass if you know the force and the acceleration.

Let's take this a little further. Assume you have several bowling balls of different mass. Some are heavy and some light. To get around problems of friction and gravity we'll assume you are an astronaut somewhere out in space. Don't worry about falling if you're not standing on anything. Unless somebody pushes you,

you'll just sit there. Anyway, let's assume you have the bowling balls out there, all in a line with the heaviest at one end of the line and the lightest at the other (see figure 5.3). Now, using the same amount of force, push on each of them. What happens? As you no doubt expect, the lightest one quickly reaches the greatest speed in a given time, say, one minute. In fact, as you go up the line you get less speed in a given time (and therefore less acceleration) in each case.

Using Newton's law we could calculate the mass of each of the balls if we measured its acceleration. Because we are overcoming the ball's inertia by pushing it, we will call this mass its inertial mass. It's important to note here that each of the balls had a different acceleration when pushed with the same force, so after, say, one minute they all had different velocities.

This is one way of determining the mass of an object, but everyone knows there is an easier way of doing this. Just step on a scale and weigh yourself. Newton was certainly familiar with this, and he was familiar with the fact that it gave exactly the same number for mass as did his inertial method. We'll call this mass the gravitational mass. That means that inertial and gravitational mass are the same. Indeed, Roland von Eötvös of Hungary showed in 1889 that they are the same to one part in a hundred thousand, and in 1964 Robert Dicke of Princeton University took this a step further and showed that they were equivalent to one part in a hundred billion.[3] For all practical purposes, then, they are exactly the same. Should this be a surprise? It was somewhat surprising to Newton, but he didn't push the matter any further. It was a coincidence, and although he wasn't quite sure why it occurred, he didn't pursue it any further.

Something even more surprising and perhaps easier to comprehend can be seen by taking things a step further. You've no doubt seen this numerous times and never thought much about it. In the inertial experiment it took a greater force to push a heavier bowling ball to the same speed as a light one. Furthermore, we know from another law that Newton gave us, namely, the law of gravity, that in a gravitational field the gravitational pull on a large, massive object is greater than that on a light object. In fact, the more massive the object, the greater the force. But as Galileo showed many years earlier, all the bowling balls, regardless of their weight, would fall at the same rate if you dropped them. How is this possible? After all, the Earth is pulling on the more massive bowling balls with a greater force, and when you were out in space, when you pushed with a greater force you got a greater velocity.

Gravity is obviously different in some way if all objects fall at the same rate in a gravitational field, regardless of their weight. Why would they do this? The reason is that inertia pulls back on the balls with a force that is proportional to its mass. The greater the force, the greater the inertial force. In short, the inertial force and gravitational force balance so that all objects fall at the same rate.

Einstein certainly wasn't the first to discover this. Newton knew all about it, but as I said, he accepted it as a coincidence. Einstein, on the other hand, wasn't so easily convinced. The outcome of his reluctance was what he called the "Principle of Equivalence." As we will see later, it would become one of the two major cornerstones of his generalized theory. The principle states that there is no way to distinguish the motion produced by inertial forces (acceleration) from motion produced by gravitational forces. In other words, if we go back to our man in the elevator, there is no way he could ever distinguish an elevator accelerated upward at 32 feet per second per second from one sitting in a gravitational field. No matter how many experiments he performed, the results would always be the same. In particular, the experiments could be mechanical or optical; it would make no difference. Also critical here is Einstein's assertion that the accelerated elevator isn't creating an "artificial" gravitational field. To him it was creating a "real" gravitational field. The two situations were exactly the same. Well . . . almost. We'll talk about the slight difference later. Incidentally, another way of stating the Principle of Equivalence is by saying that gravitational mass is equal to inertial mass.

It was through this Principle of Equivalence that Einstein was able to generalize his major postulate of special relativity, namely, that all motion in the universe is relative. In short, when you pass someone (in empty space) you have no way of knowing if you are moving, or if he is moving. In the accelerated elevator you have no way of knowing if you are accelerating, or if the entire universe is accelerating relative to you. Both cases give the same final outcome.

There is, indeed, a well-known situation where it is virtually impossible to distinguish inertia and gravity. Pilots encounter it frequently, but fortunately they have instruments to get around the problem. If a pilot is in a dense fog he cannot distinguish the gravitational pull on his body from an inertial pull when he goes around a sharp curve. It is therefore impossible for him to distinguish up and down, and needless to say if he didn't have instruments his minutes would be numbered.

EINSTEIN AS A PROFESSOR

Einstein's breakthrough to the Principle of Equivalence occurred in 1907, and strangely for several years after that, he made no further progress toward his generalized theory. As we will see, it took him ten more years to complete the theory. He remained in the patent office in Bern for two more years, but finally in 1909 he went to the University of Zurich.

Just before he began his new teaching position, he attended his first scientific convention in Salzburg. It would be his first face-to-face meeting with some of the greatest scientists of Europe, and he was curious to see them. Einstein's reputation had now spread, however, and many of them were equally curious about him. One of the people he met was Max Planck of Germany, who would play an important role in his life over the next few years.

In October he took up his new post at the University. He would be a professor "extraordinary." With the word "extraordinary" attached to it, it might seem that he would be more than a full professor, but that wasn't the case. He was still in the lower ranks, and still commanded a rather meager wage. He hoped to find time to work on his research, but he soon found that his heavy load left little time. He was determined to show that he was a good lecturer and spent considerable time preparing lectures. But the truth is, he still preferred to lecture casually and answer questions rather than lecture formally. He would usually come to class with a small scrap of paper in his hand with a list of the topics on it that he intended to talk about. But frequently he would forget what he was going to say and start talking about something that was of more interest to him. Unlike most professors he was very personable; he liked to talk to students rather than lecture to them, and he took great pains in making sure they understood what he said. They would frequently follow him to a cafe after lectures where he would smoke his pipe while he explained things to them. And occasionally he would even take them to his home, much to Mileva's dismay.

One aspect of the job he didn't like was that he was placed in charge of the laboratory. Every time he stepped into it he was sure he was going to break something. In addition, it took a lot of his time. As much as he enjoyed certain aspects of teaching, the time it took frustrated him. He wasn't getting nearly as much research done as he wanted. His extension of special relativity had gone nowhere, and he was also interested in several other problems in atomic physics and radiation theory, and he had done little on them.

While he struggled with lectures and labs in Zurich, however, his fame continued to spread. His 1905 papers, and the few that he had published since, were becoming better and better known. In particular, there was now considerable interest in his special theory of relativity. Finally he got an offer from the German University in Prague. He would be made a full professor and would be given a considerable increase in wages. The offer was to his liking, but he was uncertain about Prague, and somewhat reluctant to leave Zurich. Still, it was too good an offer to refuse; he knew he would have to consider it seriously.

When news of the offer finally got out, many of his students were disappointed, and as a result some of them and a number of the faculty signed a petition and presented it to officials at the university trying to force them to make Einstein a better offer. It did have a positive effect: they raised his wages. But they refused to match the Prague offer. They would not raise him to a full professor. After all, he was only thrity-two years old, and they wouldn't match the wages that Prague offered. So Einstein left.[4]

PROFESSOR IN A STRANGE LAND

In March 1911, Einstein and his family arrived in Prague. In many ways living conditions were better. He had a larger apartment than in Zurich, and he now had electric lights. Furthermore, his wages were high enough that Mileva could hire a maid to help her keep house. But many things were not as agreeable as they had been in Switzerland. The water was so bad that Einstein had to buy bottled water. The streets were not as clean, and the scenery less impressive. Furthermore, to Einstein's dismay the students seemed much less serious than those in Switzerland. They didn't appear to be as dedicated, and were less interested in what he had to say.

Another serious difficulty was the people of Prague. The population was divided into three groups: the Czechs, the Germans, and the Jews. And each of the groups stuck generally to themselves, and each looked down on the others. The largest group by far was the Czechs, and most people therefore spoke Czechoslovakian. Einstein did not speak the language, and therefore, to some degree, was a foreigner. (He, of course, was required only to speak German as he was in the German university.)

Einstein, ever an optimist however, ignored the problems as much as possible and made the most of it. He was always jovial and outgoing and soon had made several good friends at the university. Mileva, on the other hand, kept to herself and made few friends. Furthermore, she was quite jealous of most of Einstein's friends, and particularly jealous when he talked to other women. She became increasingly moody and complained a lot, and he in turn began avoiding her. Although few people noticed it at this stage, it was the beginning of the end of their marriage.

One thing pleased him, however. His teaching load was not as great as it had been at the University of Zurich and he had more time for research. Furthermore, the university had a large and well-equipped library. His office at the university was also large, and had a window looking down on beautiful gardens—the gardens of the insane asylum next door. He would occasionally take visitors to his office over to the window and show them inmates wandering along the paths. "They're happy," he would say. "They're not worried about quantum theory."[5]

With his new freedom, Einstein was finally able to get back to his extension of the special theory of relativity. He now had the basis of the theory, and he knew what he wanted, but he wasn't quite sure how to get it. He had formulated the Principle of Equivalence and had used it to make a number of predictions. One of them was the bending of light near a massive object. The beam from a star, for example, would bend as it grazed the sun. The reason for this is easy to see if we use the Principle of Equivalence. Consider the man in the elevator again. Assume the elevator is far out in space, well away from any gravitational influences. The man would, naturally, think he was in space, or in free-fall on Earth. If he drilled a small hole in one side of the elevator and allowed the beam from a star to cross it, he would see that it went straight across (see figure 5.4). Let's assume he marked the spot where it hit.

Now suppose that a cable is attached to the top of the elevator and it is pulled upward with increasing velocity; in other words, it is accelerated. As the beam crosses the elevator it will be deflected downward just as a bullet would be if it crossed an upward accelerating room (see figure 5.5). But by the Principle of Equivalence, the accelerating elevator is equivalent to a gravitational field. This means that the beam of light would be deflected by a gravitational field. In fact, the greater the field, the greater the deflection.

Using a "thought experiment" similar to that above, Einstein was also able to show that clocks in different gravitational fields would run at different rates. We will talk more about this later.[6]

One of the main differences in Einstein's approach to the problem now was his use of Minkowski's four-dimensional formulation, with time being the fourth dimension. When Einstein first saw the change Minkowski had made he was not impressed, mostly no doubt because he still disliked Minkowski. Einstein thought that Minkowski was just complicating things, but as he studied his formulation in detail he realized it was important, and he began taking advantage of it.

At this point Einstein was finally beginning to realize how hard the problem was going to be. It would be a much more diffi-

Fig. 5.4.
Light beam
crossing a
stationary
(or uniformly
moving)
elevator.

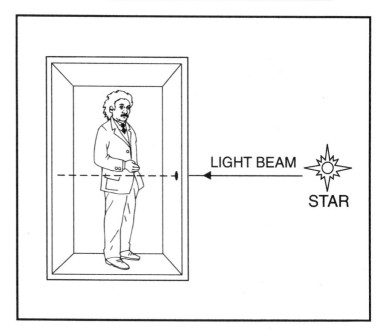

Fig. 5.5.
Light beam
crossing an
elevator that is
accelerated
upward.

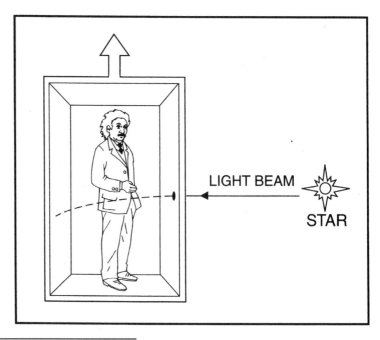

cult than the formulation of special relativity. One of the major problems was geometry. In special relativity you could use Euclid's geometry. This is the geometry that he had learned in high school (and you no doubt did, too), where only one line can be drawn parallel to an existing straight line, the interior angles of a triangle add up to 180 degrees, and so on.[7]

Einstein now realized this geometry would not work in his new theory. He needed something more general. Also, the equations in his new theory would have to have a form that was independent of the coordinate system. This was so important to Einstein that he set it up as a new principle called the Principle of General Covariance. Don't be alarmed by the big name; it's nothing more than the statement that the equations would have to be expressed in a way that put all space-time coordinate systems on the same footing.

Einstein, unfortunately, didn't know how to do this. So he talked to a mathematician friend by the name of Georg Pick about it, and Pick told him there was a branch of mathematics in which it was possible. But it was a difficult form of mathematics, and very complex. It's not known whether Einstein took him up on it, and there's no indication that he looked into it further at that time.

Einstein also had a further problem at Prague. He was, to a large degree, isolated. In formulating special relativity he had had Besso to talk to. Although Besso was no physicist he understood enough to be helpful, and Einstein could use him as a sounding board for his ideas. Einstein had always liked to discuss his ideas with somebody. He had, in fact, brought an assistant with him from Zurich, a man by the name of Ludwig Hopf. But Hopf disliked Prague and soon left. Despite the absence of an understanding ear, Einstein did manage to publish two papers on the subject while he was in Prague.

But within a year or so Einstein began to get restless. Some of the problems were getting to him, and he wasn't making as much progress on his new theory as he wanted. So when a letter came to him from his old friend Marcel Grossman inquiring about the possibility of him coming back to Zurich, he was thrilled. Grossman was now the dean of mathematics and physics at the Zurich Polytechnic—Einstein's old Alma Mater. The Polytechnic had changed significantly since Einstein had left it. Doctorates were now offered and there was considerable interest in improving science at the institute. Furthermore, his old physics professor Heinrich Weber was now dead, and Hermann Minkowski had left for Göttingen.

Einstein was delighted at the prospect of going back to his old alma mater. He would be a full professor and his wages would be as high as those in Prague. Interestingly, while Einstein waited for the offer to be finalized he got several other offers: one from the University of Utrecht, one from the University of Leiden, and one from the University of Vienna. Furthermore, Columbia University in the United States invited him to give a series of lectures on his new theory. But his only serious interest was the position in Zurich, so he turned down everything else.[8]

BACK TO ZURICH

In January of 1912 Einstein arrived in Zurich. He was glad to be back in Switzerland, but Mileva was even more delighted. She loved Switzerland, and had grown to hate Prague. One of the first things he did when he got back was visit Grossman. "Marcel, I'll go mad," he said to him almost immediately. "You've got to help me."[9] Einstein explained his new theory to Grossman and told him about the problems he had encountered. In particular he told him about the invariant form he would need for his equations, and the problems of geometry.

Luckily, Grossman was an expert in geometry. He had, in fact, done his doctoral thesis in the area. And although he wasn't an expert in the type of geometry Einstein needed, he was familiar with it. Like Pick, Grossman told him about the geometry; it had been formulated a few years earlier by Georg Riemann, Tullio Levi-Civita, and Carl Gauss. He reminded Einstein that they had even taken a course together on some of the elementary aspects of the geometry associated with the new mathematics. The mathematics was called "tensor calculus," and Grossman warned Einstein that it was extremely complex and difficult.[10]

Grossman had come to his rescue again. He even offered to help Einstein with the mathematics, but he told him he would have nothing to do with the physics. As a student he had taken several physics courses, and although he had always done well in them, he had reservations about physics. He was a mathematician, and would stick to mathematics. Einstein agreed.

THE BIG PUSH

Einstein now had an ally, somebody to talk to, and he knew what he had to do. But first he had to learn the new tensor calculus, and with Grossman's help he soon mastered it. The foundations of his new theory were in good shape. He had formulated two principles that it must satisfy: The Principle of Equivalence and the Principle of Covariance. He now needed a basic equation that would define his theory and he had several things to guide him in its formulation. In a weak field approximation it would have to reduce to Newton's gravitational field equation. Also it would have to predict the well-known anomaly in Mercury's orbit (Mercury's orbit deviated from Newton's prediction), and finally it would have to predict the bending of light around gravitating objects.

The two men struggled with the problem for months. "Compared with this problem the original relativity was child's play," Einstein wrote shortly after beginning work with Grossman."[11] Ironically, they actually looked at the equation that Einstein would show three years later was the correct equation, but they would discard it, mistakenly thinking that it didn't give Newton's equation as a first approximation (for weak fields). It seemed that they had tried every conceivable equation. Einstein finally got so frustrated he decided to abandon the Principle of Covariance. It seemed too restrictive. But he did so with a heavy heart.

With Grossman's help, Einstein eventually came up with an equation, and together they published two papers. But Einstein wasn't satisfied. The theory was not covariant and there were other difficulties: it didn't predict the anomaly in Mercury's orbit.

While Einstein and Grossman struggled, Max Planck and

Walther Nernst of the University of Berlin were considering how they could entice him away from Zurich. They had been given the task of recruiting faculty for a new expansion at the university and a new institution that was to be built in connection with it—the Kaiser Wilhelm Institute of Research. Both men realized that Einstein was now one of the leading researchers in theoretical physics; furthermore, he was young and full of promise. They decided to make him an offer he couldn't refuse. He would be made full professor at the prestigious University of Berlin, his wage would be increased, and he would be made director of the theoretical research division of the Kaiser Wilhelm Institute. As far as Einstein was concerned, though, the most interesting part of the offer was that he would have considerable freedom. He could teach or do research as he pleased.

Einstein was pleased with the offer, but worried. He felt like a "prized hen." With such an offer he would be expected to produce more "golden eggs," as he called them, and he wasn't sure that he could. Furthermore, he had never had good feelings for the Germans. After all, he had left Germany when he was young and had renounced his German citizenship because of his distaste for German militarism and discipline. Would they ask him to become a German again? As it turned out they didn't. To further complicate things, Mileva was devastated when she heard of the offer. Einstein's mother was in Berlin and she had always had a poor relationship with her. She loved Zurich and hated to leave it.

Fig. 5.6. "They want me to lay another golden egg."

Red and White Roses

When Max Planck and Walther Nernst of the University of Berlin traveled to Zurich in the summer of 1913 to offer Einstein a professorship at the University of Berlin, he refused to make his decision immediately. Planck and Nernst attended a function in a nearby town while he made up his mind. With his usual sense of humor Einstein said he would meet the train the next day, and if his answer was yes he would be wearing a red rose in his lapel. If it was no he would be wearing a white rose. When they arrived Einstein came forward to greet them; he was wearing a red rose.

Something else no doubt had some influence on Einstein's decision. His cousin Elsa was in Berlin. She had recently been divorced and had two daughters. Einstein had corresponded with her earlier, but had recently cut it off, fearing Mileva's jealousy.

TO GERMANY

Germany was on the verge of World War I when Einstein arrived in Berlin in the spring of 1914. Mileva had already traveled to Berlin with their two sons. It was clear from the beginning, however, that she hated it and within a month of arriving she packed up their two sons and went back to Zurich. It was the end of their marriage.[12]

Einstein was soon engrossed in his work. It had now become an obsession with him. He was so close, but there were still difficulties. He frequently forgot to eat as he pored over page after page of mathematics, struggling to find the right equation out of hundreds of possibilities. He talked to almost no one during this time. His cousin Elsa visited him occasionally and brought him food in an attempt to make sure he ate properly, but she knew better than to interrupt him for long periods of time.

Week after week he continued his struggle. Finally he came back to an equation he and Grossman had considered three years earlier and looked at it more carefully. He saw now that it did give Newton's equations as a first approximation and it was covariant. Furthermore, it predicted the bending of light rays around a gravitating object and explained the anomaly in Mercury's orbit. Einstein was ecstatic. He presented his results to the Prussian Academy at their next meeting.[13] Planck had long been skeptical of Einstein's attempt to formulate a theory of gravity, but he was now pleasantly surprised with his success.

In early 1916 Einstein published his results in *Annalen der Physiks*. He titled the article, "The Formulation of the General Theory of Relativity."[14] Near the beginning of the paper he made the statement, "The laws of physics must be of such a nature that they apply to systems of reference in any kind of motion." And he had, indeed, shown that this was the case. He had successfully extended special relativity to non-uniform motion. It was clear to him that the paper, which was written in the new and strange language of tensors, would be incomprehensible to most scientists, so he began with a long and detailed explanation of tensors. He then presented his new field equations for gravity and showed that they reduced to Newton's equations in the proper limit. He also showed that they satisfied conservation laws such as the conservation of energy, how they predicted the bending of light beams in gravitational fields and the behavior of clocks, and finally he explained the anomaly in Mercury's orbit.

Few people understood the details of the theory, and as expected many people scoffed at it as nonsense, particularly because it seemed so different. But as the details were studied, scientists began to realize it was an incredible theory. It had elegance and beauty, and indeed, it is still considered to be the most elegant theory that has ever been devised. Furthermore, unlike most theories, it is, for the most part, the work of a single man.

A Performing Monkey

Einstein eventually became so famous he could easily have made a large amount of money. He was offered large sums to appear on the screen for only ten or fifteen minutes. All he would have to do, they said, is stand at a blackboard with a piece of chalk in his hands. Einstein would have nothing to do with it. "Why would I want to be a performing monkey?" he said.

6

Gravity and Curved Space-Time

E INSTEIN WAS WALKING ON CLOUD number nine for weeks after he completed his general theory of relativity. "It was the greatest satisfaction of my life," he wrote.[1] The theory had not yet been tested, but Einstein had tremendous faith in it. Nobody was going to tell him it was wrong. He knew it was right.

Many scientists, on the other hand, weren't quite sure what to make of the theory at first. It was extremely mathematical and difficult to understand; furthermore, some bizarre concepts had been introduced. Many people shook their heads in dismay, but nevertheless they were in awe of Einstein. Strangely, though, although Einstein had set up the framework of the theory, got the equations he wanted, and showed that they reduced to

Newton's equations in a first approximation, he had not obtained a solution to them. This was to be accomplished by an astronomer by the name of Karl Schwarzschild. Although Schwarzschild was the director of the Astrophysical Observatory in Potsdam, he was far from his observatory when he read about Einstein's new theory. He was on the Russian front with shells bursting over his head. With all his scientific accomplishments he could easily have avoided serving in the army but he had enlisted of his own free will. By this time conditions at the front were so bad that he had contracted a rare metabolic disorder. Einstein's paper gave him a lift. He was so fascinated with it that despite his condition, the extreme weather, and the distraction of the war around him, he managed to find a solution to Einstein's equation. His solution gave the curvature of space-time around a massive object in space, a star, for example. He sent the result to Einstein.

Einstein was surprised and pleased to get the letter. He wrote Schwarzchild back immediately, saying, "I have read your letter with utmost interest. I had not expected that one could formulate the exact solution of the problem in such a simple way."[2] Within days Einstein had communicated the solution to the Prussian Academy. But Schwarzschild was not finished. He had obtained an "exterior" solution for the region around the mass, and now wanted an "interior" solution—a solution for the region inside a star, for example, and within weeks he had it. Again he sent it to Einstein. His health was now so bad, however, that he never lived to hear of its acceptance or of the glory that the solution would eventually bring him. Two months later, in May 1916, he died. He was 41.

Schwarzschild's solution was strange and certain aspects of it bothered Einstein, but he said little about them until much later. We will see, however, that they had a dramatic effect on physics.

WHAT IS GRAVITY?

Einstein had presented a new theory of gravity to the world, and we know that he abhorred Newton's idea that gravity was a force. So what was gravity in his theory? As strange as it might seem, it

> ### Einstein Can't Figure
>
> Einstein was boarding a streetcar in Berlin one day when he noticed, mistakenly, that the conductor had given him the wrong change. The conductor quickly recounted it and thrust it back at him blurting, "It is correct—the trouble is you don't know how to figure."
>
> The conductor obviously didn't know he was talking to the man who had done some of the most complicated and abstract "figuring" the world had ever seen.

was curved or warped space. How could space be curved? Space is nothing; it's just the void, the emptiness around us. And it is true, you could never hope to see this curvature. But there is a case where you can easily see a similar curvature. Take a sheet of paper, hold it up in front of you, and bend it. It's easy to see that the surface you are looking at is curved. We can see it because we live in three dimensions. To see the curvature of a three-dimensional space we would therefore have to live in four dimensions (a four-dimensional space, not four-dimensional space-time). The best we can do is to try to imagine what it might be like, by thinking about a sort of clear jelly with a twist or warp in it. I'll admit, though, that this isn't very satisfying. But even if we can't see the curvature, we can deal with it using mathematics. What we need is a geometry that describes it.

THE GEOMETRY OF SPACE-TIME

Most adults no doubt remember the geometry they took in high school. It was the geometry of flat space, called Euclidean geometry after its founder, Euclid. The foundation, or basis, of this geometry is a number of self-evident truths, or axioms. There are five in all, but one of them was controversial. It is: Through a point not on a given straight line only one line can be drawn parallel to a given straight line. Interestingly it was eventually proved that this statement is equivalent to: The sum of the interior angles of a triangle is 180 degrees.

Let's consider these two statements. It is obvious that they are valid on a flat sheet of paper. Try drawing more than one line through a point parallel to a nearby line. You'll soon find that it is impossible. Or take your protractor and measure the interior angles of a triangle. Regardless of how you measure them they'll always add up to 180 degrees (see figure 6.1). But if you try the same experiment on the surface of a large sphere you may be surprised. The sum of the interior angles will be greater than 180 degrees, and you won't be able to draw any lines through the point that is parallel to the line that is already there (see figure 6.2). If you keep the line as straight as possible it will end up being a "great circle" (a circle with diameter equal to that of the sphere) and all great circles intersect one another. The geometry on a curved surface such as this is called an elliptical geometry. It is said to be positively curved geometry.

Interestingly, this isn't the only non-Euclidean geometry. Consider the surface of a saddle. If you draw a triangle on it the interior angles will add up to less than 180 degrees, and if you try to draw a line through a point that is parallel to another line, you will find that you can draw an infinite number (see figure 6.3). This is called a hyperbolic geometry.

It would seem that the first of the two non-Euclidean geometries to be discovered would be the elliptical one—the geometry of a positively curved surface. After all, the surface of a basketball or beachball is more familiar to most people than the surface of a saddle. But interestingly it wasn't. Several people, in fact, looked at hyperbolic geometries before anyone thought of an elliptical geometry. The first to consider hyperbolic geometries was Johann Carl Friedrich Gauss. Born in Brunswick Germany on April 30, 1777, Gauss was a child prodigy, capable of incredible feats of mathematical calculation even before he went to school. He could add long columns of numbers with lightning speed and to the amazement of his teachers he proved extremely difficult mathe-

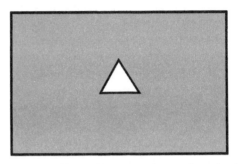

Fig. 6.1. Triangle in flat space (sum of angles is 180°).

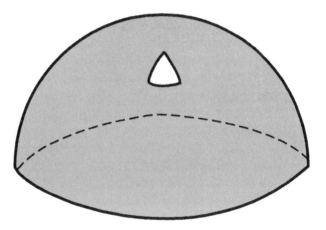

Fig. 6.2. Triangle in positively curved space (sum of angles is >180°).

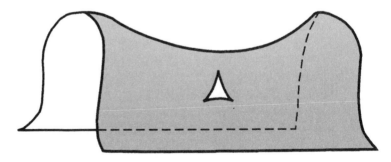

Fig. 6.3. Triangle in negatively curved space (sum of angles is <180°).

matical theorems in elementary school. His attention was soon drawn to celestial objects, and he calculated the orbit of the asteroid Ceres from only a few observations after it was lost. He then told astronomers where to find it, and indeed they found it where he said it was. At university he continued to impress his professors with his feats and was soon publishing impressive mathematical papers.

Although Gauss discovered the first non-Euclidean geometry—the hyperbolic geometry—he did not publish his results. It was not until after others had discovered it that he pointed out that he had discovered it years earlier.

Janos Bolyai, a Hungarian mathematician, made the discovery in 1823. He was the son of a well-known mathematician who was a good friend of Gauss. Janos showed considerable mathematical ability when young but he decided to pursue a career in the military. While in the military he became an excellent swordsman, once challenging a dozen of his fellow officers to a duel, one after the other. He defeated them all. In his spare time he worked out the ideas for a hyperbolic geometry and showed it to his father. His father didn't know what to make of them so he sent them to Gauss. Gauss replied that the work was brilliant, but he mentioned that he had arrived at the same results several years earlier. Janos was dejected and decided not to publish. In 1831, however, his father included them in a mathematical book he was publishing. But unknown to Bolyai the same results had by this time been published a few years earlier by Nikolai Lobachevski of Russia.

The other non-Euclidean geometry, namely, elliptical geometry, was discovered by Georg Friedrich Riemann, a German mathematician. Born in Hanover, Germany, on September 17, 1826, Riemann was the son of a Lutheran minister. He originally planned on following in his father's footsteps and becoming a minister, but when he discovered his talent for mathematics he began to have second thoughts. By the time he went to university he had decided to pursue a career in mathematics. His first encounter with non-Euclidean geometry came at his doctoral thesis defense. He presented three topics for his lecture, one of which was non-Euclidean geometry. Not expecting it to be requested he was unprepared when his committee asked for it. He had to put something together quickly, and indeed he did, and in the process he discovered elliptical geometry. Gauss was one of his committee members and was thoroughly impressed with the young scholar. Over the next few years Riemann continued to make further discoveries in mathematics but his career was cut short by tuberculous. He died in 1866 at the young age of thirty-nine.

With Riemann's discovery there were now three different geometries: Euclid's flat space geometry, Bolyai and Lobatchevski's geometry of negatively curved surfaces, and Riemann's geometry of positively curved surfaces. Riemann, however, went further than the others: he extended his geometry to three dimensions, then to four and even more dimensions. Furthermore, he considered the possibility of spaces with varying curvature.

SPACES WITH VARYING CURVATURE

Euclid's geometry is always the same. It is the geometry of a flat surface and does not vary from place to place. With either Riemannian or Bolyai-Lobatchevskian geometry we have another possibility. The curvature can be uniform throughout the space or it can vary from point to point. Riemann was particularly interested in varying curvature. The easiest way to keep track of varying curvature is to use Pythagorus's theorem. In flat space we know that the sum of the interior angles is 180 degrees, but in a positively curved space it is greater and in a negatively curved space, less. With this in mind let's consider a surface such as the neck of a vase (see figure 6.4). It is both positively and negatively curved. If we draw a large number of small triangles on it and determine the sum of their interior angles, we can keep track of the regions where it is positively curved and where it is negatively curved, and we can see the variation of curvature from place to place. This could also, of course, be done in three-dimensions, and even more. It would be more difficult to visualize, but mathematically it wouldn't be a problem. As it turns out, in space we do encounter situations where the curvature varies.

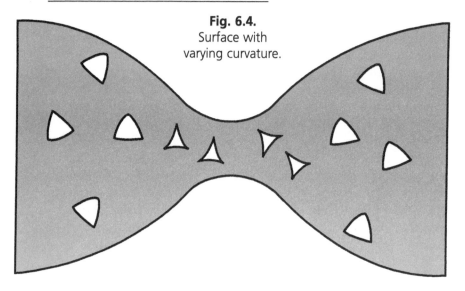

Fig. 6.4.
Surface with
varying curvature.

EINSTEIN'S GRAVITY

Einstein showed that matter curves space; the greater the mass, the greater the curvature. The easiest way to visualize this is to picture a thin rubber sheet, stretched so it is tight, like a trampoline. If we place a Styrofoam ball on this sheet it will dent it slightly. If, on the other hand, we place a wooden ball of about the same size on it, it will dent it more, and finally if we place an iron ball of about the same size on it, it will practically fall through.

Using one of these balls, say, the wooden one, and the rubber sheet, we can get a good idea of how Einstein pictured the sun and the planets. The rubber sheet represents the space around the sun. When the ball is placed on it, it warps the sheet, just as the space around the sun is warped by the sun's mass. If you now place a marble on this curved sheet it will roll immediately into the wooden ball, just as any object placed in space near the sun would fall into it. If you give the marble a slight push, however, you can get it going around the ball.

The marble is like a planet orbiting the sun. Note that it is not "forced" into the sun when it is placed on the sheet, as it would be

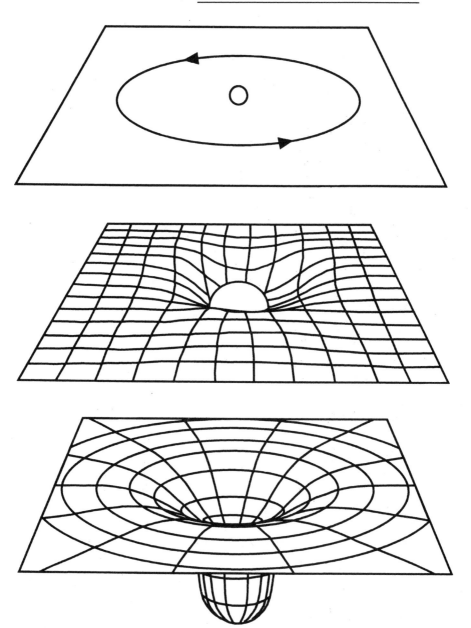

Fig. 6.5. The curvature of space due to matter (the sun or a star).
The bottom diagram shows extreme curvature.

according to Newton's theory of gravity. It falls toward it because of the curvature of the space (sheet) in the region around the sun.

The path that a planet takes around the sun in Einstein's theory is called a "geodesic." In mathematical terms this is the shortest distance between two points. It is also the longest, but we'll ignore that for now. Hold on, you say. How can the planets follow geodesics if they are going around the sun in elliptical orbits? An ellipse is curved. Indeed, it is curved in ordinary space, but in Einstein's theory we're dealing with four-dimensional space-time, not space itself, which is three-dimensional, and in space-time the orbit of a planet is indeed straight.

Space-time in Einstein's theory is non-Euclidean, and therefore a non-Euclidean geometry is needed to describe it. In this non-Euclidean space, objects move in geodesics—in other words, they take the shortest path between two points. But what is straight in space-time appears as a curve to us as we are in ordinary space. In short, then, according to Einstein's theory, matter (such as a star) curves the space around it and all objects that move through this space move in geodesics. These geodesics appear to us as elliptical orbits.

It's important to point out that objects in a gravitational field are still obeying Newton's first law, but now it is in a modified form (i.e., all objects in uniform motion continue to move uniformly unless they are acted upon by a force).

TIDAL FORCES

Everything we have talked about so far has been determined using the Principle of Equivalence, but as I mentioned earlier there is a slight flaw in the way we stated this principle. Let's go back to the elevator again. If it is accelerated upward at 32 feet per second per second, you would think you're in a gravitational field on the earth. But if we look carefully at the earth's gravitational field, we notice something strange. Gravity acts directly toward the center of the earth and this means that if we draw a picture of the field using lines to represent it, we will have converging lines as shown in figure 6.6.

But if the lines converge they can't be parallel as they would

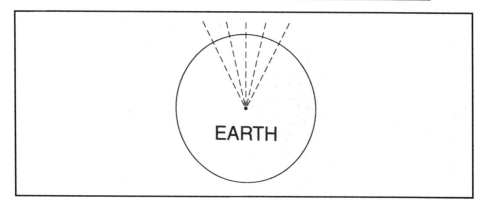

Fig. 6.6. Earth's gravitational field. Dotted lines are field lines of gravity. Note that they are directed toward the center of the earth.

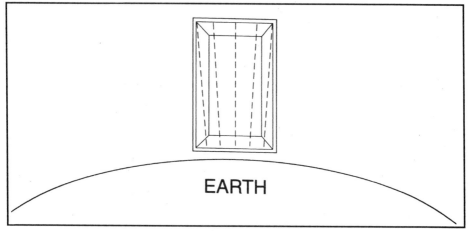

Fig. 6.7. Non-uniform field lines in an elevator. Field lines are not uniform throughout the elevator. They are slightly closer together at the bottom.

have to be if the field was uniform. Indeed, they tell us that the field is not uniform (see figure 6.7). But when the elevator is accelerating upward out in space (far from a gravitational field), the lines that represent the "field" within it will be uniform (see figure 6.8). This means that the Principle of Equivalence is only valid when the elevator is infinitely small. In practice, however, the convergence is so small for an object as large as the earth that to a good approximation there is an equivalence for something as small as an elevator.

Let's look a little closer at the effect of this non-uniform field. Assume that you are an astronaut falling toward the earth in this field. What would happen? The gravitational pull on your feet would be greater than that on your head, and you would be stretched as you fell. What I should have said is that the space curvature is greater near your feet than near your head, but the effect is the same. You would be in an region of varying curvature, and it would tend to pull you apart.

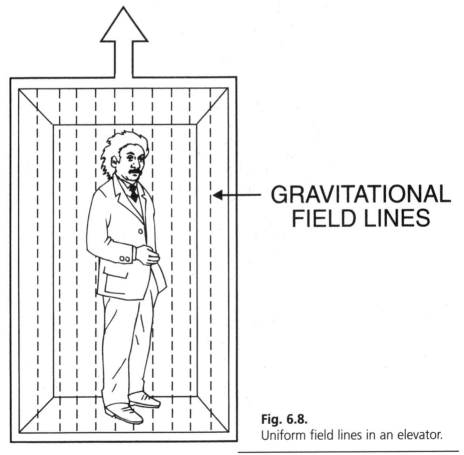

GRAVITATIONAL FIELD LINES

Fig. 6.8.
Uniform field lines in an elevator.

For the earth this effect is small, but if you were falling toward a small, dense planet, or star, you could be pulled into something resembling a piece of string. Scientists refer to these forces as tidal forces. The name comes from the fact that forces such as these cause the oceanic tides on Earth.

How do they cause tides? We know that the gravitational pull between two objects depends on their separation. But if one of the objects (or both) is large, some of its parts will be much closer to the second object than its other parts. Consider the earth and moon. The greatest pull from the moon will be at C, and the least at D. Since the earth is made up mostly of ocean, water will flow toward C. At the same time it will be left at D. The result will be tides as shown in figure 6.9.

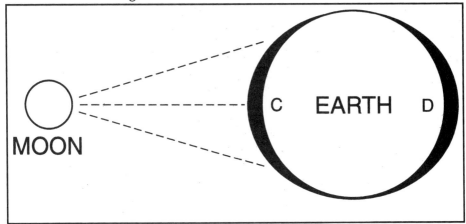

Fig. 6.9. Tidal pull of the moon on Earth. Note that the pull is greatest on side C.

WAVES OF GRAVITY

If a massive object curves space, one of the first questions that come to mind is: How fast does it act? In other words, does it act instantaneously or does it take a certain amount of time for the warping to reach a particular point in space? To make my point a little clearer, consider an object falling to the surface of the earth. The distribution of the earth's mass will, naturally, change as a result of this fall, and this in turn will produce a small change in the earth's gravitational field (or in relativistic terms, a small change in the curvature of space-time around the earth). Granted, the change will be exceedingly small, but there will, nevertheless, be a change. Is it possible that this change takes place instantaneously? That's impossible according to special relativity which states that nothing can move with a speed greater than that of light. But the gravitational field of the earth must be signaled that a change in the earth's mass distribution has occurred, and a change in the field must propagate outward from the earth to effect the change. This disturbance is now known as a gravitational wave. Such waves were postulated by Einstein shortly after he formulated his general theory of relativity. In particular, he showed that they would propagate with the speed of light.

Einstein found that gravitational waves would be given off by matter when it is subjected to acceleration. In an attempt to determine the magnitude of these waves he considered a rotating rod, but soon found that the intensity would be far too low to be measurable. He therefore made the calculation for a binary system made up of two stars orbiting one another. Again, he found that for ordinary stars the intensity of the waves would be too low to be measurable. Only in the case of very small, massive stars would there be any chance of measuring the gravitational waves emitted.

Gravitational waves are, in many ways, analogous to electromagnetic waves, the waves responsible for radio and television. Electromagnetic waves are generated by accelerating electric charges—electrons, for example—racing back and forth on an antenna. In the same way, gravitational waves are produced by masses that race back and forth (accelerate), but compared to electromagnetic waves they are exceedingly weak.

Joseph Weber of the University of Maryland began looking for evidence of gravitational waves in the late 1950s. He was convinced that his best chance of detecting them was to look for them in space—perhaps given off from a dense binary star system. He designed a detector in the form of a large cylinder. When a gravitational wave passed through this cylinder, it would vibrate with a certain frequency (that of the gravitational wave). Gauges mounted on the side of the cylinder that were extremely sensitive to small changes in the cylinder would allow Weber to detect the wave.

Weber reported success in 1969 but when other scientists tried to verify his results they couldn't. It was soon clear that Weber had been mistaken. But he did generate a lot of interest. New, larger detectors were built over the next few years, but the waves were still not detected.

In 1974, however, a breakthrough came. The waves weren't detected directly, but they were detected indirectly. A binary star system consisting of two very dense stars was discovered by Joseph Taylor and Russell Hulse of the University of Massachusetts using the gigantic thousand foot radio telescope at Aricebo, Puerto Rico. It is now referred to as the binary pulsar system, since at least one of the two stars in it is believed to be a very dense star called a pulsar. The two stars in the system are very close and they orbit one another in about eight hours. For a stellar system this is an extremely short period of time. Any system of this type radiates energy—both electromagnetic and gravitational—and as a result it gradually slows down. In essence, its period changes.

Taylor and several colleagues carefully monitored the system for several years, looking for any sign of a slowdown and in 1978 they announced that they had detected one. Comparing it with the expected slowdown they found that it could only have slowed down by the amount that it did if it was giving off gravitational waves. This wasn't a direct detection but it was exceedingly strong evidence for their existence, and as such was very important.

Since that time several other, more sensitive experiments have been set up, but so far gravitational waves still have not been detected directly. A new project called LIGO, short for Laser Interferometer Gravitational Wave Observatory, is presently being built near Hartford, Washington, and at Livingston, Louisiana, and should be operational sometime soon. A joint French-Italian team is building a similar facility near Pisa, Italy. It is called VIRGO. When operational, VIRGO and LIGO can work together as a team. When they become operational most scientist are convinced that we will finally detect gravitational waves directly.[3]

GRAVITATIONAL LENSES

Another important prediction that came out of Einstein's general theory of relativity is known as the gravitational lens. We mentioned earlier that a gravitating body deflects light; a light ray grazing our sun, for example, will be deflected so that we see it a short distance away in the sky from where it really is. Could this effect be used as "lens" if the conditions were right? Indeed, it was predicted that it could shortly after Einstein put forward his theory. A nearby star, for example, could be used to magnify a distant object in the sky, say, a galaxy. The star would, of course have to be positioned in exactly the right place in line with the galaxy. The probability of this actually happening is actually quite low. But if the alignment is close, it was shown that you would get two or more images (usually distorted) of the galaxy (see figure 6.10).

In May 1979 three astronomers, D. Walsh, R. Carswell, and R. J. Wyman announced that they had detected what might be the first gravitational lens.[4] They discovered two side-by-side quasars that appeared to be the same and had exactly the same physical properties. A second similar object was discovered by R. J. Wyman in 1980. Astronomers are now convinced that these are indeed gravi-

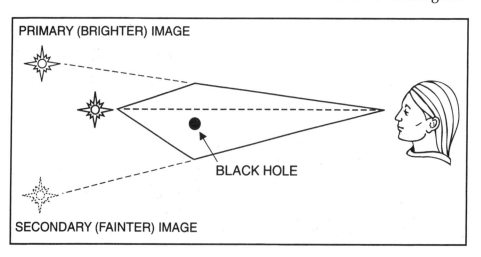

Fig. 6.10. A gravitational lens
(see chapter 9 for a description of a black hole).

tational lenses and that we are, in each case, seeing more than one image of a single quasar.

THE TWIN PARADOX REVISITED

Earlier we mentioned the problem of twins, where one of the twins takes a trip in a spaceship to a nearby star and back, and the other twin stays on earth. Depending on the speed of the spaceship there can be a considerable difference in the ages of the twins when they get back together again. Indeed, if one of the twins travels at very close to the speed of light (of the order of 99.999 percent), this difference can be hundreds of years.

The paradox, as it existed before general relativity, was that you couldn't be sure which of the twins would be the youngest when they got back together after the flight. Either could argue that they were the one that moved off into space, since all motion is relative according to special relativity. But when Einstein formulated his general theory of relativity he showed that there was, indeed, a, distinct difference between the twins. Earlier we called them Bob and Fred. Bob experienced a force on his body as a result of acceleration, and Fred did not. In fact, Bob had to accelerate to get away from the earth, then decelerate as he approached his destination star. Then he had to accelerate when he returned to Earth and finally when he arrived at Earth he had to decelerate. According to Einstein he would be the younger of the two when they got back together.

The twin paradox has caused more controversy than any other aspect of general relativity. It is very difficult for most people to believe that twins could actually age at different rates. Many people refused to believe it at first, questioning whether this was really a "physiological" time difference. Perhaps it was just a mathematical time difference. We now know that it is indeed a physiological difference and one twin would be older than the other. Don't get your hopes up, though. Relativistic time travel can never be a "fountain of youth." Remember, the twin that is younger notices nothing strange in his clock; time runs at the normal rate for him, just as it does for the other twin. If he lived to 70 it would seem like 70 normal years to him.

You might ask how far we can go with this kind of thing. We are, of course, restricted by the speed of light, but if one of the twins moves at a speed very close (and I mean *very* close) to that of light, millions of years could pass back on Earth. Time is, indeed, much stranger than we might have thought.

7 Testing the Theory

YEARS BEFORE HE FINALIZED HIS GENERAL theory of relativity, Einstein was already thinking about how it could be tested. He soon realized that a beam of light passing near a massive body would be deflected by its gravity, or more properly, by its space curvature. The beam from a star would be ideal. There were always some stars near the limb of the sun; unfortunately the only time you could see them was during an eclipse. Einstein made the calculation: the stars would be displaced during an eclipse by .87 seconds of arc (the angle between the original position of the star and its new position). He then talked to astronomers about the test, but no one seemed interested until he met a young Berlin astronomer by the name of Erwin Freundlich. Freundlich offered to organize an expedition to the next eclipse, which would be visible in Siberia in the summer of 1914.

Freundlich and several assistants left for Siberia about a month before the eclipse, and on August 1, 1914, while they were setting up their equipment, Germany declared war on Russia. Freundlich and his assistants were taken prisoner. Fortunately they were not detained for long. Germany had also taken prisoners and an exchange was soon arranged.

Einstein was disappointed, but ironically it was a lucky break for him. At this time, unknown to him, he was predicting the same displacement that Newton's theory gave. Years later, when he had completed his theory, he found that the displacement should be double .87, or 1.74. If Freundlich and his assistants had measured double his prediction, it likely would have gotten little attention.

EDDINGTON AND THE 1919 ECLIPSE

It would be another five years before anyone would try again to verify Einstein's prediction. Outside of Germany few knew anything about Einstein's theory. World War I was on and almost no scientific information was being exchanged between Germany and most of the rest of the world. One of the few that had access to German journals was an astronomer in Holland by the name of Willem de Sitter. De Sitter became interested in Einstein's theory and passed details of it to scientists in England. Over the next few months he wrote several articles on it for the British journal of the Royal Astronomical Society, *Monthly Notices*. Arthur Eddington, a professor of astronomy at Cambridge University and Director of the Cambridge Observatory, read the new theory with interest. He was, in fact, one of the few in England capable of understanding the mathematics of the theory. A child prodigy, he had distinguished himself at Cambridge by ranking first in mathematics. After studying the theory in detail he was so convinced of its validity he gave little thought to tests that might prove it. The theory was so elegant and beautiful it had to be right. But when the Astronomer Royal, Sir Frank Dyson, pointed out that the bending of starlight around the sun could be tested in an eclipse that would take place in May, 1919, Eddington was eager to help. The eclipse would, in fact, be an ideal one. It would take place near the Hyades cluster in the constellation of Hyades, which meant that a large number of stars would appear near the sun during the eclipse.

Dyson and Eddington made preparations for the eclipse expedition. Strangely, when they began, England was still at war with Germany. Few knew of the plans at this time. Testing a theory that a German had put forward would no doubt have been unpopular, but Eddington was a pacifist and Einstein was not actually a German (he was a Swiss citizen).

Two sites were chosen for the expedition: Principe Island in the Gulf of Guinea and Sobral in Brazil. Eddington was head of the expedition to Principe, A. C. D. Crommelin, the one to Sobral. Two months before the eclipse Eddington and his assistants set out for Principe, allowing plenty of time to set up and test their equipment. For several months Principe had been under a drought so there seemed to be little danger of clouds or rain. But when the day of the eclipse arrived, Eddington woke to rain pounding on his tent. He was devastated, but decided to make the best of it anyway. By midmorning the rain had stopped, but it was still cloudy. The eclipse would occur at midday. Even though all they could see was a bright disk through the dense clouds, preparations went ahead. To Eddington's delight as the eclipse began the clouds started to break up. He and his assistants began taking pictures. Their main interest was the stars around the disk of the sun, not the eclipse itself. The first ten plates showed no stars, but as they began the last six, there was a sudden break in the clouds and a few stars were visible. But it wasn't until the last plate that they got a good distribution of stars.

Eddington worried. Would it be enough? He was so anxious he began developing the plates the next evening and indeed in the last plates stars were visible. This region of the sky had been photographed back in England with the same telescope several months before they left and Eddington had brought the plates with him, so a comparison could be made. He realized that conditions were inadequate for a good measurement; nevertheless, he went ahead.

There was a displacement, and it appeared to be close to Einstein's prediction. Eddington was overjoyed, but he was cautious. He knew that better measurements would have to be made.

When he arrived back in England he used more delicate instruments to make the measurements. The first few gave a displacement close to .87 arc seconds, the value predicted by Newton's theory. Eddington panicked but soon saw that there was an error due to the lenses he was using. He continued measuring and the next plates gave a value very close to Einstein's prediction, but not exactly equal to it. Still, under the poor conditions that the plates had been taken they were within experimental error. Furthermore, there was still the Sobral plates. The Sobral group had no problems with the weather and had got several good plates, but they had been delayed. They had to get a good photograph of the sky in Hyades under dark sky conditions (no moon) with which they could compare their eclipse photographs. This took two months. Then there were further delays on their trip back, so it was five months before they were back in England.

Eddington couldn't wait to make measurements from their plates. He assisted them, and within a few weeks the Sobral plates were shown to give an average displacement of 1.98 arc seconds. Eddington had got an average of 1.61 arc seconds from his plates, and the two results, when averaged, gave 1.79. Einstein's prediction was 1.74 arc seconds. Within experimental error, therefore, the prediction had been verified.

Einstein knew of the expeditions and eagerly awaited the results, but because of the difficulty of communication between England and Germany, he had a long wait. In June he finally heard from his friend Ehrenfest that the expedition had been successful. Still, he did not know if his theory had been verified. Upon hearing the news he wrote to his mother, "It is said in a Dutch paper that both expeditions obtained successful photographs of the solar eclipse, so that the results should be known within 6 weeks."[1]

The results, however, did not come in six weeks, and Einstein soon became anxious. In September he wrote to his friend Ehrenfest in Holland asking if anything was known. The answer came on September 22. "Eddington found star dislocation at sun's rim provisional . . . between .9 and 1.8."[2] This was not particularly good news, but it was based on early measurements. Einstein, however, wasn't worried; he was still confident.

On Thursday, November 6, 1919, a joint meeting of the Royal Society and the Royal Astronomical Society was held in London. There was considerable anticipation as everyone present knew that the results of the two expeditions were going to be announced. The philosopher Alfted Whitehead later said, "There was an atmosphere of tense interest that was exactly that of a Greek drama."[3]

Part of the tension came from the fact that many in the audience considered the debate to be between Newton and the "German," Einstein. This was far from Eddington's view, however; from the beginning he had been strongly biased in favor of Einstein. The well-known physicist, Sir Oliver Lodge, on the other hand, had publicly scoffed at the new theory, and had made a substantial bet that it would not be vindicated.

Sir Frank Dyson was the first to speak. He described the expeditions, the difficulties evaluating the plates, then announced to a hushed audience: "After careful study of the plates I am prepared to say that there can be no doubt that they confirm Einstein's prediction. A very definite result has been obtained that light is deflected in accordance with Einstein's law of gravity."[4]

Eddington then got up and discussed the Principe results; he was followed by Crommelin, who discussed the Sobral results. Then came the most dramatic statement of the evening from Nobel Laureate J. J. Thomson, the chairman and president of the Royal Society. "This is the most important result obtained in connection with the theory of gravity since Newton's day. It is one of the highest achievements of human thought."[5]

What was perhaps ironic is that almost no one in the hall understood the details of the new theory. It was just too complex mathematically to be understood by anyone without a strong mathematical background. As Eddington left the assembly, one of the other scientists stopped him. "There's a rumor that only three people in the entire world understand Einstein's theory," he said. "You must be one of them." Eddington looked at him in silence for several seconds. "Don't be modest, Eddington," the man said.

The Photograph

When Einstein got the photographs from the English eclipse expedition he stared at them for the longest time in awe. "It's marvelous, truly marvelous," he said.

Thinking he was pleased with the results of the expedition his wife said, "You must be very pleased."

"Yes," said Einstein, "but look at the quality of these photographs. I never thought photography had reached such perfection."[8]

Eddington shrugged. "Not at all," he said. "I was wondering who the third might be."[6]

The next day, November 7, the headlines in the *London Times* were, "Revolution in Science—New Theory of the Universe—Newton's Ideas Overthrown." According to the article, "our conception of the fabric of the universe must be fundamentally altered." But, strangely, Einstein himself was so unknown in England that there were no details about him. No one knew who he was.

News of the event reached America even before it reached Germany. But again, no one had ever heard of Einstein. A brief report in the *New York Times* summarized the story from the *London Times*. The following day a much longer article appeared, and soon the story spread across the nation. Over the next few months hundreds of articles appeared, written for the most part by people who knew nothing about the theory or the man behind it.

When Einstein received a telegram telling him of the announcement he was pleased. But even then his confidence showed. A student that he showed the telegram to asked, "What would you have done if it had not been confirmed?"

Einstein replied: "In that case I'd have felt sorry for the dear Lord, because the theory is correct."[7]

Einstein was already fairly well-known throughout Germany, but with the confirmation of his theory he was suddenly a celebrity. Hundred of reporters were soon besieging him. He was so overwhelmed he didn't know what to do.

HYPERSPACE

The mathematics of Einstein's theory is, indeed, complex but there are simple ways of explaining his prediction of a bending of light around the sun. We saw earlier that a light beam crossing an accelerating elevator out in space would be deflected downward slightly, and by the Principle of Equivalence the same deflection would be expected in a gravitational field. But we can look at this from a different point of view. According to Einstein's theory the beam is deflected because the space around the sun is curved (see figure 7.1). And we know that we can represent this curvature using a ball and a rubber sheet. Scientists do something similar to this. They imagine themselves taking a "slice" through the space-time region around the sun. If we took a slice through ordinary flat space it would give us nothing more than a flat sheet. But if we took a slice through the curved space-time near the sun we would get a curved surface like the one shown in figure 7.1. It is referred to as a hypersurface.

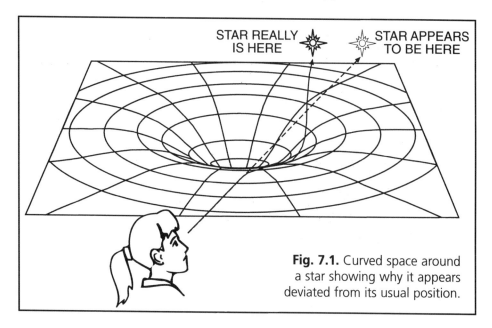

STAR REALLY IS HERE   STAR APPEARS TO BE HERE

Fig. 7.1. Curved space around a star showing why it appears deviated from its usual position.

Now, let's consider a light beam traveling across this surface close to the sun. It is easy to see that the beam will be deflected, and we will therefore see stars slightly deflected from their usual positions.

THE PRECESSION OF MERCURY'S ORBIT

Another of Einstein's predictions was a slow change of the orientation (a "precession") of Mercury's orbit. All the orbits of the planets are elliptical, but Mercury's is the most elliptical so the effect is easiest to measure in its case. Any ellipse has a major axis—the line drawn through its center (see figure 7.2). If the ellipse precesses, however, the direction in which the line points will gradually change. Indeed, in time (a very long time—200,000 years) the line will trace out 360 degrees.

Einstein knew when he was formulating his theory that it would have to explain things that Newton's did not. Though Newton's theory was an excellent one, there appeared to be a few things that it couldn't explain. One of the few was a slight anomaly that had been observed in Mercury's orbit. In short, Mercury wasn't following Newton's prediction exactly. But there could be several reasons for this; one of the most obvious was that it was being perturbed by a nearby planet. Perturbations of this type are common in the solar system. Every planet is perturbed by every other planet. Both Neptune and Pluto were, in fact, discovered as a result of their perturbations: Neptune as a result of a perturbation on Uranus and Pluto as a result of a perturbation on Neptune.

For many years astronomers believed that Mercury was being perturbed by a planet inside its orbit. They had actually named the planet Vulcan, after several astronomers had announced that they had found it.[9] In each case, however, the announcement turned out to be false. Vulcan was elusive. Was it possible that something else was causing the deviation? Einstein was convinced that there was.

The amount of the deviation from Newton's prediction was exceedingly small, but it could easily be measured by astronomers. Einstein's theory predicted a precession of about 43 seconds of arc

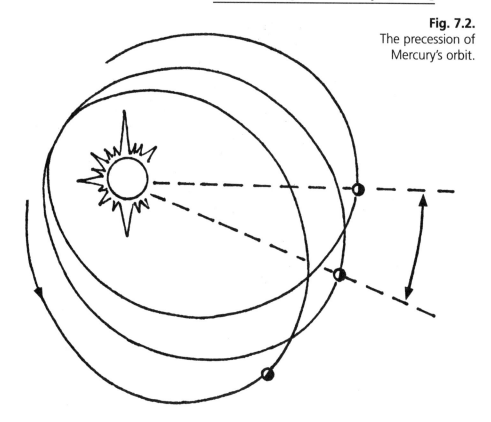

Fig. 7.2.
The precession of
Mercury's orbit.

per century, and this agreed within experimental error with the predicted deviation.

A dramatic verification of the precessional prediction came soon after Joseph Taylor and Russell Hulse discovered the Binary Pulsar.[10] As I mentioned earlier, it consists of two very dense objects orbiting rapidly around one another. Because they are so close to one another, precession in the system is rapid and can easily be measured. And indeed, when it was measured it satisfied Einstein's theory exactly.

CLOCKS IN A GRAVITATIONAL FIELD

Another verification of Einstein's theory came from the prediction that clocks in different gravitational fields run at different rates. A clock in a strong field, for example, runs slower than one in a weak field. Why this is so is easily understood if we go back to our accelerated elevator. In this case, however, we will need two elevators, with one accelerating faster than the other. By the Equivalence Principle they will have different "gravitational fields" with the field of the one that has the greatest acceleration being the strongest. Suppose we want to compare clocks in these two elevators. The only way we could do this is send a light signal from one of the elevators (call the observer Mary and assume she is in the slower of the two) to the other (the observer in the other elevator is Gloria). Assume that at the end of each second a pulse goes out from Mary's elevator toward Gloria's. But Gloria's elevator is accelerating at a greater rate and the pulse will therefore be "compressed" by the time it gets there. Gloria will therefore think that Mary's clock is running fast compared to hers, and since Gloria is in a higher gravitational field (by the Principle of Equivalence), clocks in stronger gravitational fields will therefore run slower (see figure 7.3).

This means that a clock on the surface of the earth will run slower than one at the top of a high building. The difference, however, is extremely small. Even if we took the clock at the top of the building to outer space the difference would be too small to measure. What we need for a measurable difference is a gravitational field that changes a lot over a relatively short distance. In the 1920s such an object was discovered; it was only slightly larger than Earth, yet it had the mass of the sun. Called a white dwarf, it had a gravitational field that changed significantly over a short distance.[11] Because it was very close to the bright star Sirius (it was actually in orbit around it), it was called Sirius B. When its density (weight per unit volume) was measured, it was found to be 61,000 times that of water. Using Sirius B, astronomers were able to check Einstein's prediction that clocks in high gravitational fields run slower than those in weak fields.

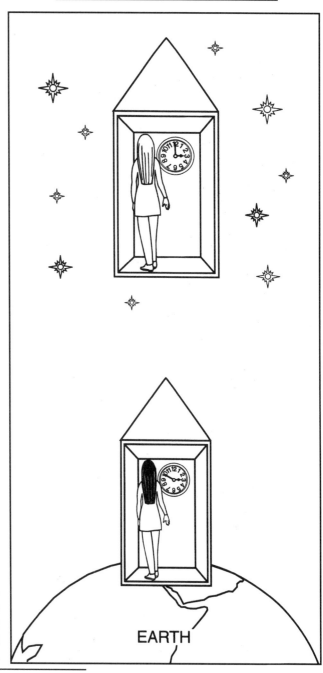

Fig. 7.3.
A clock on the
surface of Earth
runs slower than
the one above Earth.

EARTH

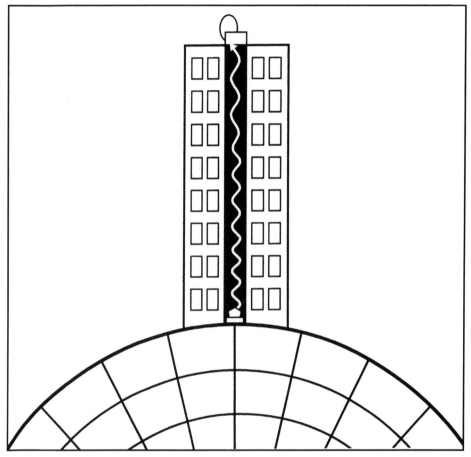

Fig. 7.4. Cobalt emitter is at the bottom of the tower. The absorber is at the top.

To see how this is possible, it's important to remember that anything that vibrates is a clock; this means that a vibrating atom can be thought of as a clock. And indeed, vibrating atoms were the "clock" that was used in the experiment. Furthermore, we know that accelerating atoms give off radiation, or light. So, let's consider the vibrating light wave from a dense star such as Sirius B. As it climbs out of the strong gravitational field it will vibrate at a lower frequency. One way of thinking about this is that the tiny bundle of light will lose energy in its fight to free itself from the gravitational field, and as it does it will vibrate less vigorously. The effect is now referred to as the gravitational redshift because the light will "redden" as it loses energy. In the case of Sirius B, astronomers verified that the frequency of light emitted was reddened by the amount predicted by Einstein's theory.

Needless to say, measurements such as this are difficult to make and there are many sources of error. Because of this, astronomer Jesse Greenstein and several colleagues carefully analyzed a dozen white dwarfs in 1977. They found that the slowing of clocks in the vicinity of all of them was, indeed, in accordance with Einstein's theory.

The most accurate verification of the effect, however, came after a young German physicist, Rudolf Mössbauer, discovered that the nuclei of atoms in a crystal can emit and absorb radiation at very precise frequencies.[12] They therefore act like "tiny nuclear clocks" that keep very accurate time. The effect is now called the Mössbauer effect. In 1959 R. V. Pound and G. A. Rebka of Harvard University realized that the Mössbauer effect could be used to test the gravitational redshift predicted by general relativity. They set up the experiment in a physical lab at Harvard using radioactive cobalt. The cobalt was placed in the basement of the building and holes were drilled through several floors to the roof above, which was 74 feet up. An absorber was set up there to absorb the emitted rays (see figure 7.4). Measurement of the emitted and absorbed frequencies then allowed them to test Einstein's theory, and it was again verified.

An Einstein Quote

"For the most part I do the thing which my own nature drives me to do. It is embarrassing to earn so much respect and love for it. Arrows of hate have been shot at me, too; but they never hit me, because somehow they belong to another world, with which I have no connection whatsoever."[13]

OTHER THEORIES AND OTHER TESTS

By the 1960s several theories similar to Einstein's had been put forward. Most of them had the same features as general relativity, in other words, they assumed gravity was curved space-time, but they all had slightly different predictions. It was therefore even more critical that precise tests of the theory be made. The best known of these other theories was one by Robert Dicke of Princeton University and his student Carl Brans. Its roots go back to England in the late 1930s. The physicist Paul Adrien Maurice Dirac, one of the founders of quantum mechanics, began to wonder if the universal gravitational constant of the universe was actually a constant. Maybe it was changing so slowly that we didn't notice it. He presented a theory of the universe with a gradually varying gravitational constant, but within a few years it was shown to be flawed. The idea of a varying gravitational constant, however, stuck and in the early 1960s Dicke and Brans formulated a slight variation of general relativity using the idea. The predicted values for the deflection of light grazing the sun and the precession of Mercury's orbit were slightly less in the Brans-Dicke theory than it was in Einstein's theory.

But if observations had verified Einstein's theory, how could the Brans-Dicke theory be correct if it gave different predictions? Dicke suggested that they may not be as exact as believed. In the case of the precession of Mercury's orbit the Brans-Dicke theory gave about 5 arc seconds less than Einstein's theory. If, however, the sun was not perfectly round, but flattened at the poles because of its rotation, his prediction would be closer to observations than Einstein's theory. In 1964 he and several associates began an experiment to measure the sun's shape very precisely. He announced his results in 1967: the sun was, indeed, oblate, and as a result the precession of Mercury's orbit agreed much more closely with the prediction of his theory.

Others soon became interested in the experiment. Henry Hill of the University of Arizona was among them. He began a similar experiment soon after Dicke announced his results, and in 1974 he announced that Dicke was not correct. The sun was not oblate;

what Dicke had mistaken for oblateness was a small, slow oscillation of the sun's brightness. Others soon verified Hill's result.

TESTS USING QUASARS, SATELLITES, AND LASERS

In the 1960s astronomers discovered strange objects in space that are now called quasars. They look like ordinary stars but emit large quantities of radio waves. One of the first of these objects to be discovered was called 3C 273 (it is the 273rd object in the third Cambridge catalogue).[14] By the early 1970s astronomers were able to measure the deflection of the radio waves from a quasar as they grazed the sun, and they were therefore able to use it to check general relativity. In practice they used two quasars, 3C 279 and 3C 273, which lie close together in the sky. As it turns out, on October 8 of each year 3C 279 is eclipsed by the sun. In 1970 astronomers carefully measured the angle between the two quasars as 3C 279 approached the sun and was eclipsed. Einstein's theory was again verified.[15]

In another test the small laser beam reflector left on the moon by the Apollo 11 astronauts was used. Extremely accurate measurements of the distance to the moon at any time can be made using this reflector, and since the Brans-Dicke theory predicts a small change in the universal gravitational constant, the distance to the moon should be gradually changing. I should mention that other things such as tides also affect the distance, so corrections had to be made. But again when the final results were in, Einstein's theory was favored over the Brans-Dicke theory.

With the advent of atomic clocks, scientists were able to measure very brief time intervals very accurately. Atomic clocks were therefore a natural for testing the theory. In 1971 atomic clocks were placed in a jet plane. The plane was flown around the earth from east-to-west, then from west-to-east. Time dilations as a result of the plane's speed (as predicted by special relativity) and gravity at various altitudes (as predicted by general relativity) were checked by comparing the clocks to similar clocks at the Naval Observatory in Washington, D.C. Both predictions of Einstein's theory were verified within experimental error.[16] Similar tests have been done more recently using satellites in orbit and they have also verified Einstein's theory.

ANTIGRAVITY

In the 1930s the British theorist P. A. M. Dirac predicted the existence of an "antiparticle" to the electron, namely, a particle with the same mass but opposite charge. Since the electron was negatively charged, this new particle would be positively charged. This positive electron, or "positron," was discovered by Carl Anderson in 1933 at the California Institute of Technology. We now know that there are antiparticles corresponding to all particles. Indeed, when a particle and an antiparticle collide they annihilate one another with the release of considerable energy. Nowadays we can easily create antiparticles in large accelerators.

The earth and solar system are, of course, composed of matter, but it is possible that some parts of the universe are composed of antimatter. If true, we would have to ask ourselves how antimatter would affect general relativity. In other words, does general relativity explain antimatter?

Matter, of course, attracts other matter gravitationally, but it would repel antimatter. Therefore, if we had a beam of antimatter on Earth, it would be deflected upward slightly by gravity. Indeed, if we could make a "ball" of antimatter and drop it, it would be repelled and rise, rather than fall to the earth.

In Einstein's theory we deal with accelerating elevators that "simulate" gravity. A ball composed of matter would fall just as it does on Earth. Unfortunately, we have no way within general relativity of explaining how a ball of antimatter would rise in the elevator. The theory seems to be incapable of dealing with antimatter. Is this a flaw? Not necessarily. So far we haven't actually observed objects of antimatter in the universe; we see only antiparticles coming in as cosmic rays, and there may be a reason for this.

EINSTEIN THE CELEBRITY

Soon after the November announcement in 1919, Einstein was known throughout the world for his new theory of gravitation. Few people understood it and there was still considerable opposition, mostly from people that didn't understand it. Sir Oliver Lodge referred to it as "repugnant to commonsense." Dr. Thomas See of the University of Chicago wrote, "The Einstein theory is a fallacy. The theory that the 'ether' does not exist and that gravity is not a force but a property of space, can only be described as a

Fig, 7.5.
Reporters besieging Einstein.

A Happy Fella

Despite the honors and praise showered on him, Einstein had a distaste for fame and fortune. He shunned publicity, and from all accounts was happy.

He liked to be comfortable and frequently wore old crumpled clothes—and no socks. He shaved in the bathtub with the same soap be used for bathing.

Having different kinds of soap when one would do was just too complicated, he said.

crazy vagary, a disgrace to our age." Engineer George Gillette referred to the theory as "cross-eyed physics . . . utterly mad . . . pure drivel and voodoo nonsense."[17]

Einstein heard many of the rumors, but he did not let them bother him. Despite the outrage on the part of some, he had a tremendous amount of support and he was, without question, now the world's most famous scientist. He was soon besieged with strange offers, almost as if he were some sort of freak. Many people did, indeed, regard him as supernatural. It was all strange and new to him. The publicity confused him but he took it in stride, posing good-naturedly for photographers and artists.

But as the years passed, interest in general relativity gradually

What am I?

While he was still in Berlin an English reporter asked him to describe himself.

He replied: "Today I am considered in Germany as a German scientist and in England as a Swiss Jew, but if I one day become a *persona non gratis* I will be a Swiss Jew in Germany and a German scientist in England."[18]

began to fade. The differences from Newton's theory were small and Newton's theory was much less complex mathematically. Few people studied the theory in detail. Then astrophysicists began to find strange new objects in space—bizarre objects that could be explained only by general relativity.

Black Holes and Other Exotic Objects

S CHWARSZCHILD'S SOLUTION OF THE FIELD EQUA-
tions of general relativity bothered Einstein. It had
a "singularity" at the center of the mass. This is a
region where the theory breaks down and is no
longer valid. But this was expected. A singularity
also occurs here in Newton's theory. What was unexpected was another singularity at a finite distance
from the center of the mass. Not only did the solution
break down here, but it cut off the region inside it.
Indeed, Schwarzschild also noticed it and calculated
the radius at which it occurred for the sun, finding it
to be about three kilometers. This is small compared
to the overall radius of the sun, nevertheless it was
important. What was the significance of this region?
It seemed as if there as a "hole" in the center of the
curved space—a hole we had no access to.

In 1916 Ludwig Flamm looked at the geometry of the region near the cutoff. He found there was a "funnel" leading up to the radius that cut us off. This radius is now referred to as the gravitational radius (see figure 8.1).

What did Einstein think of it? He no doubt noticed it early on, but he said little about it for several years. Finally in 1922 the French mathematician Jacques Hadamard cornered him at a conference in Paris in 1922 and asked him about it. Einstein was hesitant, and it was obvious that he had worried about it: "It would be a true disaster for the theory," he said. "And it would be difficult to say . . . what would happen physically because the formula does not apply any more."[1]

The only way around the problem, it seemed, was to prove that no mass could exist inside the gravitational radius, and this was the view that Einstein took. Eddington also took notice of the region, referring to it as the "magic circle." It was magic in that it seemed to be cut off from the outside world. Although Eddington thought seriously about the problem, he was rather whimsical about it in his popular books. There he referred to it as a region where "the mass would produce so much curvature of the space-time that space would close up around the star." He was, of course, exaggerating. Space does not "close up" around the star, but as we will see it is indeed a bizarre region.[2]

Could something actually exist inside the magic circle? Both Eddington and Einstein were adamant that it couldn't. It was a region to be ignored as far as they were concerned.

Strangely, a similar phenomenon had been noticed almost two hundred years earlier. In 1783 John Michell, the naturalist and Rector of Thornhill in Yorkshire, England, presented a paper to the Royal Society of London. He reported that there might be objects in space that we wouldn't be able to see. The velocity of light was well-known at the time, and it was also known that any gravitating body, such as the earth, had an escape velocity. This was the velocity required to completely escape its gravitational pull. Consider, for example, a rocket blasting off from the earth. With a medium thrust it will rise, then fall back to earth. If the thrust is increased it will eventually go into orbit around the earth, and if it

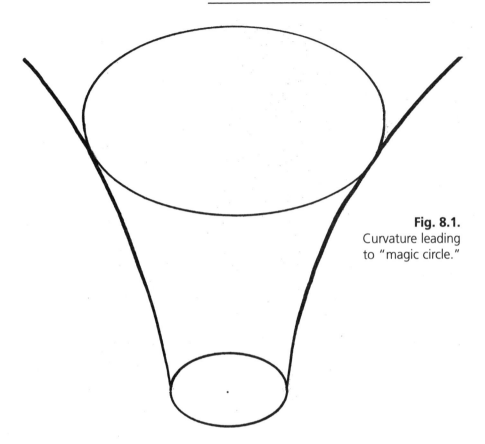

Fig. 8.1.
Curvature leading to "magic circle."

is increased a little more it will completely escape. This is referred to as the escape velocity. The greater the gravitational field, the greater this velocity. The sun, for example, will have a much higher escape velocity than the earth. In the case of the earth, this velocity is known to be about 25,000 miles per hour.

Now, let's place the earth where the sun is. We're doing this so we have a reference system around it. Then, keeping the density (mass per unit volume) constant, assume that the earth expands. As it expands, its mass increases and as a result its gravity also increases. This means that as it gets bigger, the escape velocity from its surface increases. By the time it gets to Mercury it will have increased significantly. Assume that it continues growing, past Venus, Earth, and finally it approaches Mars. Michell found that as the surface of this "huge Earth" approached Mars the escape velocity from its surface approached the speed of light. Indeed, if it went a little further the escape velocity would be greater than that of light, and since light would no longer be able to leave its surface we would not be able to see it.

The French natural philosopher and mathematician Pierre Simon Laplace noticed the same phenomenon slightly after Michell. He discussed it in his famous book *Systems of the World*.[3] Interestingly, when the third edition of the book came out, it was no longer included. Laplace had no doubt become suspicious of the object and decided not to include it.

TUNNELS IN SPACE

Many years passed with Einstein saying no more about the region. But he must have been troubled by it, for in 1935 he teamed up with Nathan Rosen to examine the geometry around a very massive point, as Flamm had done many years before. Like Flamm they followed the curvature of the space down to the gravitational radius. It was funnel-shaped and appeared to end at the gravitational radius. But as Einstein looked at the solution more carefully he saw that there was a "mirror-image" solution on the other end of the funnel. In other words, there was a "tunnel" through space with the center region cut off (see figure 8.2). Einstein wondered where this tunnel led. The only answer seemed to be "another universe." The idea was repugnant to him. Checking to see if it was possible to get through it he found that a speed greater than that of light would be needed, and of course by special relativity, this was impossible. He was relieved.

A few years earlier in 1932, Georges Lemaître of Belgium discovered something particularly strange about Schwarzschild's singularity. He showed that it wasn't a true singularity after all. He included the result in a paper on cosmology and published it in an obscure journal, so few people saw it. One that did, however, was H. P. Robertson of Harvard University. Using Lemaître's result, he showed that the mass inside the gravitational radius was not inaccessible after all. News of this reached Einstein in the late 1930s and he was alarmed. It wasn't possible; he was sure the region inside the gravitational radius was inaccessible. It had to be! If it wasn't, it could give rise to serious problems for his theory. He therefore set out to prove that it was. He considered a cluster of particles (they could be stars) that attracted one another gravitationally. The overall object was spherical and the particles orbited the center.

What would happen if this cluster got smaller and smaller? In particular, what if it was forced inside its gravitational radius? Einstein made the calculations. As expected the particles moved faster and faster as the cluster contracted. They obeyed the same laws that planets do and we know that the inner planets travel faster in their orbits than the outer ones. Indeed, Einstein found that at one and a half times the gravitational radius the particles would reach

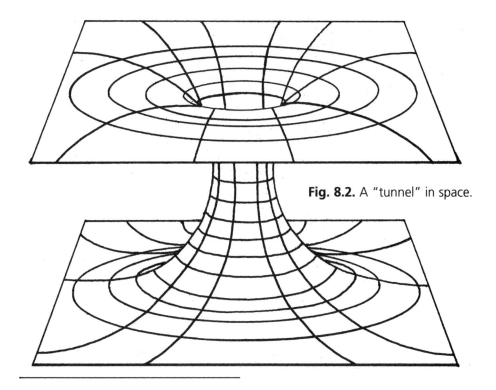

Fig. 8.2. A "tunnel" in space.

the speed of light. To go any further they would have to have a speed greater than that of light and this was impossible. Einstein was relieved. Mass couldn't exist inside the gravitational radius because it would have to be traveling at speeds greater than that of light. The region could therefore be ignored.

But Einstein had missed something. He had assumed the particles exerted an outward force as a result of their motion (or perhaps a gas pressure) and it was this force that was stopping the object from collapsing beyond a certain radius. What he didn't consider was that a powerful implosion could overcome gravity. Where would such a powerful implosion come from? As we will see, a star eventually collapses in on itself, producing a very powerful implosion.[4]

OPPIE AND HIS CRONIES

In the same year (1939) that Einstein was trying to prove that there was no access to the Schwarzschild radius, Robert Oppenheimer and his students at Berkeley, California, were showing that there was. Oppenheimer had many areas of interest. He was one of the first to bring the new science of "quantum mechanics" from Europe to America, but he was also fascinated by Einstein's theory of general relativity and he had read about some interesting work that a young Indian physicist by the name of Subrahmanyan Chandrasekhar had done on collapsing stars.

Born in New York in 1904, Oppenheimer's intelligence was obvious from the beginning. Without having to work, he was always at the top of his class. But he was far from lazy and even when very young he had a tremendous thirst for knowledge, and he was an avid reader.

In 1922 he entered Harvard University where he majored in chemistry. He completed the four-year course in three years with little effort, but was still unsure at this stage what he wanted to do. In his last year, however, he took a physics course from Percy Bridgeman. He enjoyed it so much he decided he would switch to physics for graduate work. He applied to Cambridge University in

England, hoping to work under the famous physicist, Ernest Rutherford. He was accepted but was assigned to G. P. Thomson instead of Rutherford. Thomson, an experimentalist, stuck him in the basement making thin films of beryllium.[5]

Oppenheimer began to have second thoughts about his choice. His nature, temperament, and clumsiness were not well-suited to laboratory work. He was soon depressed and began to think about leaving. Then one day Neils Bohr visited his lab. Oppenheimer talked to Bohr at length about some of the important new discoveries that were being made in Europe. They excited him and he realized he had made a mistake. His real interest was theoretical physics and within months he left England for Göttingen, Germany. He was enthralled with his new environment; at Göttingen he worked with Paul Dirac, Max Born, and others. Of particular importance he soon became an expert in the new area of quantum mechanics.

Upon completing his Ph.D., he sailed back to the United States. He had applied for several positions and to his delight he found that he was in considerable demand. Numerous offers came in, including one from his old alma mater, Harvard. But Oppenheimer decided to accept a joint appointment at Berkeley and Caltech in California. Although he wasn't a good lecturer at first he soon improved and within a short time he had about a dozen graduate students working for him. They were more than graduate students to him, though; he took them to cafes and even home with him occasionally. They were with him so much that his colleagues eventually began referred to them as "Oppies cronies."

One of his students was George Volkoff, a Russian emigrant who had come from the University of British Columbia. Oppenheimer was interested in Chandrasekhar's work on white dwarfs. Chandrasekhar had shown that a star less than 1.4 solar masses would collapse to a white dwarf when it finally ran out of fuel, but he had said nothing about what happens to stars with a greater mass when they died. Oppenheimer wanted to find out. In particular, he wanted to know what general relativity said about these stars. He assigned the problem to Volkoff.

It is well-known that a star is balanced between two forces: the force of gravity pulling it inward and an outward pressure that results from the gas pressure of the star, which is caused by the energy generated in the "thermonuclear furnace" at the center of the star. Throughout its life these forces are equal and opposite. Eventually, though, the thermonuclear furnace begins to run out of fuel (hydrogen). What happens when this occurs? As Chandrasekhar showed, gravity overcomes the star and it begins to collapse in on itself. In the case of a small star, one less than 1.4 solar masses, the star collapses very slowly to a white dwarf. It takes millions of years for the collapse to occur.

Volkoff looked at stars with a mass greater than 1.4 solar masses. He showed that there would also be a collapse in this case, but it would be much more dramatic, occurring in a very short time. Of particular importance, the end result would be a tiny star composed mostly of neutrons—a neutron star.[6] Several years earlier Fritz Zwicky, an astronomer at Mt. Wilson Observatory in California, had predicted that such a star would form in a supernova explosion. He had speculated about it but had given no proof that they actually existed. There was considerable interest in Volkolf's result, but until it was proven that neutron stars existed it couldn't be taken seriously.

Volkoff had shown, however, that there was another limit. Neutron stars could form only from collapsing stars that had a mass between 1.4 and 3.2 solar masses. It said nothing about masses beyond 3.2 solar masses. Oppenheimer had no interest in stopping here; he wanted to know what happened to stars beyond this mass, so he assigned the problem to another of his students, Hartland Snyder. Snyder was one of the most able of Oppies cronies. He had considerable mathematical savvy, and he had a good knowledge of Einstein's theory. He attacked the problem with gusto, but the result he got surprised him, and when he showed it to Oppenheimer, Oppie didn't know what to make of it either. Snyder had shown that when a star with a mass of greater that 3.2 solar masses collapsed, it just kept collapsing. There was nothing to stop it; it collapsed forever, and it left a region of space that was cut off from us. In 1939 Oppenheimer wrote a friend, "We have been working on static and nonstatic solution for very heavy masses . . . the results have been very odd."[7]

The result was, indeed, odd. It didn't make sense, and there certainly wasn't anything in nature that resembled the end state of the collapsing star—at least not to their knowledge. There was considerable interest in the result, but the United States was soon engrossed in World War II and everyone soon forgot about it. Oppenheimer was assigned to direct the Manhattan project and gave no more thought to the problem.

When the war was over, few remembered that such a solution had been obtained, and Oppenheimer was now working on other things. There was little interest throughout the 1950s, but in the early 1960s things began to change. First, a new type of object was discovered in the depths of space. Called quasars, they gave off so much energy that there appeared to be no rational way of explaining them.[8] A number of scientists remembered Chandrasekhar and Oppenheimer's calculations and suggested that quasars might be extremely massive stars in a state of collapse. Several models were put forward but they were very speculative.

Then in 1967 another discovery startled the scientific world. Antony Hewish and Joselyn Bell in England discovered rapidly pulsating radio sources. Because they were pulsing so rapidly they had to be small; furthermore, they seemed to be quite close to us. Several models were put forward. Were they white dwarfs? It was soon shown that white dwarfs presented problems. The only thing left seemed to be neutron stars, and in 1968 Thomas Gold of Cornell showed that they had to be neutron stars.

But what about the even more bizarre objects beyond the neutron star? John Wheeler of Princeton University began referring to them as "black holes." Using computers he and his students, Kent Harrison and Masami Wakano, began examining their properties. They verified Volkoff and Snyder's solution, then went on to examine the objects in more detail. Scientists in Russia soon became interested in the strange new objects. Yakov Zel'dovich and Igor Novikov and others began a similar program in Russia, and within a few years the theory of black holes was well-developed.

PROPERTIES OF A BLACK HOLE

Let's consider the collapse of a star massive enough to end as a black hole. Its final mass has to be at least 3.2 solar masses, but it no doubt loses some mass as it collapses so the star initially probably had to be considerably more massive, say, about 8 solar masses. As it ages its fuel is used up, and eventually it begins to expand to a red giant. Then finally as it runs out of fuel it becomes unstable and gravity overwhelms it. We saw earlier that the final collapse for small stars took millions of years. In this case the collapse is almost instantaneous. In a tiny fraction of a second the star collapses in on itself. As it falls it gains speed and strange things happen to the light from the star. The particles of light that are emitted from the surface (we refer to them as photons) are caught up in the curvature of the space created by the collapsing star. As the collapse continues, the density of the core increases and the photons have to exert more energy to break free from the surface. As they lose energy they change in color. Furthermore, those leaving at an angle are forced into curved paths. Then, as the core becomes very small they take up orbits around it. Finally, beyond a certain stage the remaining photons are trapped.

At this stage the matter of the collapsing star passes through the surface at its gravitational radius. This surface is known as its "event horizon." No light can leave the region inside it, and we would therefore be unable to see it directly. We would, however, be able to see it indirectly because it would block off background stars. Because of the loss of energy it would now be black—it would be a black hole.

What about the matter that makes up the star? It has now collapsed inside the event horizon so it is impossible to see. We know, however, that it continues to collapse to the center of the black hole where it becomes a singularity. All the mass of the star is now in this singularity.

A question that naturally comes to mind is: Why does the star, as seen from a distance, appear to stop collapsing before it reaches the singularity? The answer has to do with time. Earlier we saw that time passes slower in a strong gravitational field than it does in a weak one. In fact, the stronger the field, the slower time passes. Let's assume, then, that we are in space watching the star collapse; furthermore, assume we have a telescope so that we can see a clock on its surface. As we watch it the clock runs slower and slower, and as the star approaches its gravitational radius the clock appears to stop. But if we look carefully we see that the clock on its surface never quite stops, and the surface of the star never quite reaches the gravitational radius.

There is another view of this event, however. Let's say we have someone brave enough to ride the collapsing star down. What would he see? Would he see the same thing as the external observer? It turns out that he wouldn't. His clock would appear to him to be running normally as he rode the star to its fate. He would pass through the event horizon in a short, finite time. Once inside the black hole, however, he would be in a never-never land. He would never be able to get back out through the horizon. If he tried, he would find that he needed a speed greater than that of light. The event horizon would be receding from him, and of course we can't travel at speeds greater than that of light so he is trapped. He would also notice something else strange: space and time have interchanged their roles. We have complete control over space; in other words, we can move wherever we want, but we have no control over time; it passes in the same way regardless of what we do. Inside the event horizon our observer would be pulled into the singularity, regardless of what he did; he would have no control over space, but he would have some control over time.

ARE THERE OTHER TYPES OF BLACK HOLES?

The black hole we have been discussing so far is of the type discovered by Schwarzschild. It is non-rotating and arises in the collapse of a non-rotating star. We now refer to it as a Schwarzschild black hole. Einstein showed that there was no access through the tunnel or "wormhole," as they are now called, that is associated with these black holes. A speed greater than that of light was needed to get through them.

But in nature most stars rotate. In fact, you would be hard pressed to find one that didn't spin, and of course when a star of this type collapses it continues to spin. In fact, because of the conservation of angular momentum (spin) it would spin more and more rapidly as it got smaller. A figure skater uses this principle when she pulls in her arms while doing a spin. She brings some of her mass closer to her spin axis and as a result she spins faster. By the time a spinning star ends its collapse it is spinning very rapidly.

Einstein looked for a solution to his equations for the case of a spinning star, but he didn't find one. In 1963, however, Roy Kerr of New Zealand, who was working at the University of Texas, did find one. It was complicated and the resulting black hole was much more complex than the Schwarzschild black hole. To begin with, it had two surfaces; the inner one was the usual spherical event horizon, but outside it was another surface we now refer to as the static limit. This surface was farthest from the event horizon at the equator and coincided with it at the poles (see figure 8.3).

What was the significance of the static limit? Earlier I talked about frame dragging. If you were in a rocketship and approached a spinning black hole you would be dragged around it. The closer you got, the faster you would go. The static limit is the point where you would reach the speed of light. Outside this surface you could always use your retrorockets to slow yourself down, but once inside, you would be pulled around the black hole, regardless of what you did. To stand still you would need a speed greater than that of light, and of course that's impossible.

What about the region between the static limit and the event horizon? It's called the ergosphere, and as we will see, it is a particularly interesting region. If we proceeded through the ergosphere and then through the event horizon, we would see the singularity at the center of the black hole, but it would be different from the Schwarzschild singularity; it would be a ring.

The Kerr black hole is, indeed, quite different from the Schwarzschild black hole, but the most important difference is the wormhole associated with it. After discovering his solution, Kerr examined the geometry of the space around the spinning black hole. Like the Schwarzschild black hole it had a funnel leading up

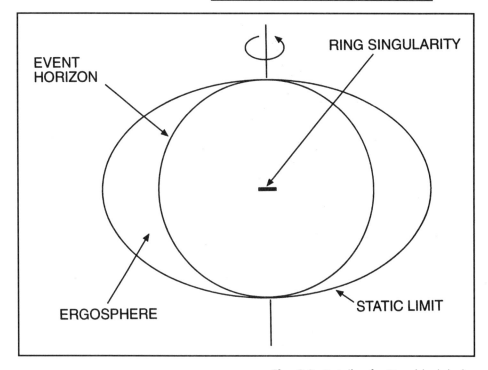

Fig. 8.3. Details of a Kerr black hole.

to the event horizon, and a mirror-image funnel on the back end. Einstein had found that a velocity greater than that of light was needed to get through the Schwarzschild wormhole. Was this also true for the Kerr wormhole? Kerr looked into this and found, to his surprise, *that it wasn't.* You could, in theory, pass through the Kerr wormhole with a speed of less than that of light. But where would you end up? Another universe was a possibility, but the most reasonable answer was a distant point in our universe. Alas! Kerr black holes could be used to travel to distant points in our universe. But don't get your hopes up. As we will see, there are many problems that have to be overcome. One thing in our favor, though, is that most, if not all, stars in the sky spin. Therefore, if a star collapses to a black hole, it will most likely be a Kerr black hole.

What about other types of black holes? To answer this we have to consider what properties manage to survive when a star collapses. Certainly the gravitational field does, as does spin, as we just saw. Are there other things? We know there is another long-ranged field like the gravitational field, namely, the electromagnetic field. It is similar to the gravitational field in many ways, arising from charge rather than mass, and it's easy to show that a black hole could have an electromagnetic field in addition to a gravitational field. If the star, for example, had a charge on its surface, this charge would survive the collapse and we would have a charged black hole with an electromagnetic field.

Fig. 8.4.
Checking the "No Hair" Theorem.

It's not too likely that this type of black hole occurs in nature. Charge is easily lost from a star; furthermore, it is neutralized when there is an opposite charge around. We know of few stars that have an excess of charge; our sun, for example, is neutral. Nevertheless, the charged black hole is a different type and it must be considered.

A solution to Einstein's equations for a charged star (or black hole) was actually obtained as early as 1916 by Hans Reissner and Gunnar Nordström of Germany and Holland respectively. Like the Kerr black hole, the charged black hole has an ergosphere, a singularity, and a wormhole that is traversible.

What about other properties that are preserved? As it turned out only these three—gravity, spin, and charge—are preserved. This was proved in what had been whimsically called the "No Hair Theorem."[9] (In essence, a black hole can't have hair.) Still, it's easy to see that with them, another type of black hole is possible, namely, one with both spin and charge. The solution for this case was found by Ted Newman and his students of the University of Pittsburgh.[10] In all, then, we have four types of black holes.

EXTRACTION OF ENERGY

An interesting result for three of the above types of black holes was obtained in England in 1971. The theorist Roger Penrose showed that energy could be extracted from them.[11] It may be a tricky process to pull off in practice, but it is feasible. Penrose showed that if a pellet is projected into the ergosphere and it breaks into two pieces, and one of the pieces falls through the event horizon to the interior of the black hole, and the other escapes, the one that escapes will come out with considerably more energy than it went in with. There will be a net gain of energy. Where does this energy come from? From the black hole itself. In the case of a Kerr black hole much of the energy is tied up in the spin. If the Penrose process is applied to it and energy is extracted, it will spin more slowly. When all the spin is finally extracted it will become a Schwarzschild black hole.

In the case of a Reissner-Nordström or charged black hole, as energy is extracted the black hole will lose charge. Finally, when all the charge is gone, it will become a Schwarzschild black hole. With a little imagination we can think of all kinds of possibilities for such a process. Small black holes could, for example, be put in orbit near Earth with some sort of device for extracting energy nearby. This energy could then be beamed to Earth. If this were possible we would have our energy needs satisfied for millions of years.

SPACE-TIME DIAGRAMS

In the chapter on special relativity we saw that space-time diagrams are extremely useful. We will see that they are also very useful in general relativity, particularly those associated with the curvature of space-time around black holes. Things are, however, a little more complicated in this case. Let's begin by looking a little more closely at the Kerr and Reissner-Nordström black holes. Previously we mentioned that they had an event horizon and a static limit. To be honest, we were keeping things simple the first time around; in reality there are two of each of these surfaces.

Consider the Kerr black hole. Let's assume we begin with a Schwarzschild black hole and add spin. Initially it will have one event horizon and we can represent it in a space-time diagram (see figure 8.5a). Now start it spinning. A second event horizon will soon be seen just above the singularity. The outer event horizon will also have moved inward slightly (see figure 8.5b). If we continue adding spin the two event horizons will continue to approach one another. Finally when the spin is very high the two event horizons will merge and become a double event horizon (see figure 8.5c). Will this double event horizon look any different from a single one? It will no doubt look the same, but there is something significantly different about it. Remember that as you cross an event horizon, space and time interchange their roles. If you cross a double event horizon they obviously interchange twice so you get no change. Space and time stay the same.

What will happen now if we continue to add spin? It turns out that the double event horizon will move inward toward the singularity, and if we add enough spin it will disappear into the singularity. What will be left? It might seem that there would be nothing, but we still have the singularity. Now, however, it isn't surrounded by an event horizon; it is "naked." Naked singularities have generated a lot of interest in the last few years. But, in reality, we still don't know much about them.

The situation in the case of a charged, or Reissner-Nordström black hole is the same. If you start with a Schwarzschild black hole and add charge you will again get two event horizons, and as you

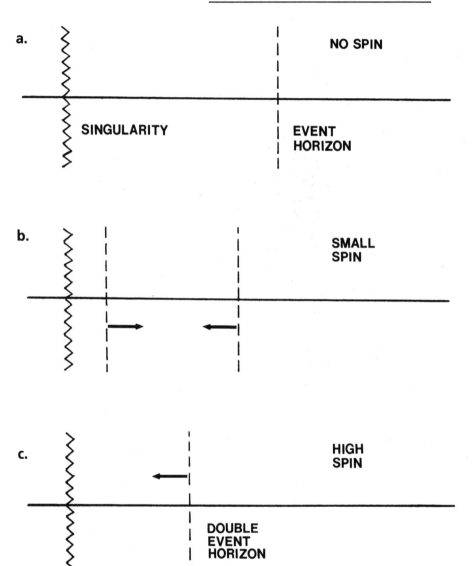

a. NO SPIN

SINGULARITY

EVENT
HORIZON

b. SMALL
SPIN

c. HIGH
SPIN

DOUBLE
EVENT
HORIZON

Fig. 8.5a–c.
Diagrams illustrating the double event horizon of a black hole.
Note that the horizons approach one another as the spin
is increased until they eventually merge.

continue to add charge they will approach one another and merge. With more charge being added the double event horizon will approach the singularity and finally, as in the previous case, we will get a naked singularity.

What we're interested in now is what would happen if we tried to explore the region around one of these black holes. In particular, what would happen if we tried to enter one? The best way to illustrate this is with the use of a space-time diagram (see figure 8.6). Let's begin with the simplest case, where we have one event horizon. We'll assume there are two astronauts, Pete and Mike, sitting in a space ship at some distance from the black hole. Mike is curious and brave enough to take a trip into the black hole. Pete is hesitant; he decides to wait behind. He'll watch from a distance. Both astronauts have clocks, of course.

As Pete watches Mike head for the black hole he sees him get closer and closer to it, but strangely he never quite seems to reach it. We can represent this in our space-time diagram as follows. The solid line AB shows Mike's trip from Pete's point of view.

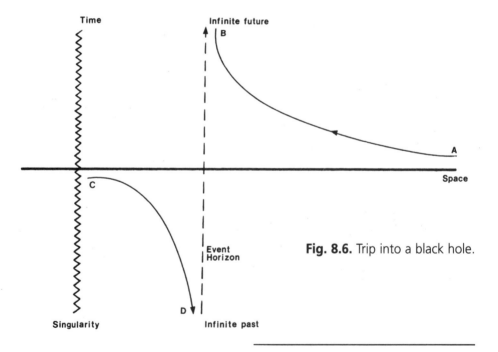

Fig. 8.6. Trip into a black hole.

Note that he gets closer and closer to the dotted line (the event horizon) but he never gets to it. Mike doesn't see it this way, however. In a relatively short period of time he passes through the event horizon and into the interior of the black hole. Let's assume now that he is near the singularity and wants to get out. His path would be represented in the space-time diagram as CD. We see that he would be moving backward in time, and would therefore go into the past. But of particular importance he would approach the event horizon closer and closer from the inside, but he would never reach it.

Scientists saw early on that there were difficulties with this diagram. It didn't tell the entire story. Something was missing, but no one was sure what to do. The critical step was taken by Martin Kruskal of Princeton University. Strangely, Kruskal wasn't an expert in general relativity. He became interested in the theory after hearing about the important advances being made in its applications and he decided on a whim to organize a study group to learn it. While studying the theory he realized that there was a transformation that would allow scientists to see the space-time around a black hole much more clearly. He took his discovery to John Wheeler. Wheeler didn't think it was important at first, but when he finally realized its significance he published it, giving Kruskal credit.

The space-time diagram that resulted from Kruskal's transformation looked quite different from the diagram we showed earlier. There are now two singularities and two event horizons. Furthermore, the two event horizons are at an angle of 45 degrees to one another (see figure 8.7).

Again, as in the space-time diagrams in special relativity, a world line at an angle of greater than 45 degrees represents an impossible trip because it is at a speed greater than that of light. Incidentally, space is again on the horizontal and time on the vertical. Three types of trips are possible within this diagram; they are represented by A, B and C. A is a trip between two points outside the black hole, and not very interesting. B is a trip through the event horizon and into the singularity. Trip C is the most interesting; in this case we pass through two event horizons and into another universe, or to a distant point in our universe. We see from

Fig. 8.7.
The Kruskal
diagram.
Three trips
from Earth
(or a point
outside a
black hole)
through the
region near
and into a
black hole
are shown.

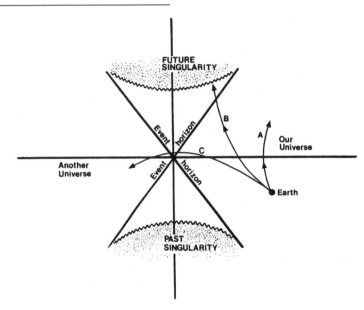

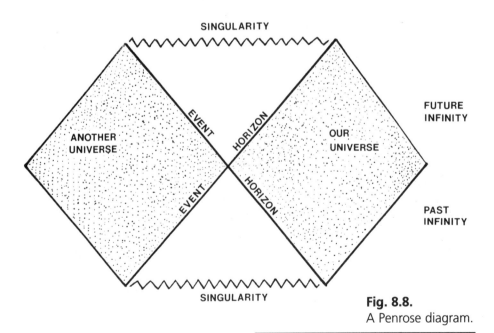

Fig. 8.8.
A Penrose diagram.

the diagram why the Kerr wormhole is accessible (it is at an angle of less than 45 degrees).

Roger Penrose showed that this diagram can be simplified slightly—at least the points at infinity can be simplified. He brought them into the diagram (see figure 8.8). Note that he has straightened out the singularity.[12]

PROBING A BLACK HOLE

Suppose now that we want to probe the region around the black hole, but we don't want to take any chances on getting too close. We could probe it with a searchlight. The space is curved so the beam will get caught up in the curvature, but how it curves will tell us a lot about the black hole and the region around it.

Assume we are at a safe distance from a Kerr black hole and we shine the beam toward it. For simplicity we'll assume we are in the plane of the equator. As we bring the beam closer and closer to the black hole it is bent more and more. Eventually, at the outer photon sphere, it will be caught in a circular orbit. As I mentioned earlier, there are two photon spheres. In this case we'll assume the beam is

Black Holes on the Internet

If you would like to find out what it would be like to orbit around a black hole, or pass through the event horizon and into the singularity, the best place to look is the Internet. There are numerous programs online related to black holes. Ted Bunn of the University of California has a site that answers questions about black holes. Martin Rees, Scientific American, Jillian, and the Royal Greenwich Observatory all have sites on black holes. Several sites feature simulations of light rays interacting with black holes and models of accretion disks around black holes. Addresses of these sites are given in the notes in the back of the book.[13]

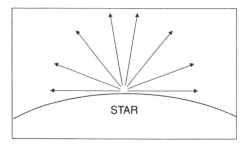

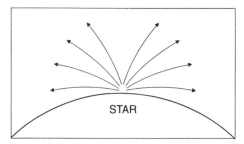

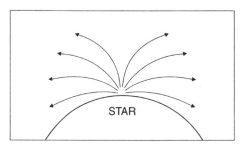

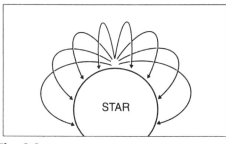

Fig. 8.9.
Light rays around a star
as it collapses.

coming in opposite to the direction of spin of the black hole. In other words, if the black hole is spinning clockwise, the beam is going around the outer photon sphere of the black hole in the counterclockwise direction.

Now switch the beam so that it will enter in the direction the hole is spinning. This time you can get it much closer to the black hole before it is captured by a photon sphere (in this case the inner photon sphere). Between these two photon spheres is a whole range of possible orbits, but only for beams that come in at an angle to the equatorial plane.

For completeness let's go back to the Schwarzschild black hole, or rather the star that collapses to the black hole. We'll assume we're on the surface as it collapses, and again we'll probe the space around it. As the collapse begins we will see that the rays are slightly deflected. Moving the beam around as the star continues to collapse, we find that it gets bent more and more. Finally, as the star passes through its photon sphere the beam gets caught up in a circular orbit around the star. Then, as we pass the event horizon, no matter where we point the beam, it does not leave the star. It is trapped (see figure 8.9).

REAL BLACK HOLES: THE CANDIDATES

So far we've discussed only theoretical black holes, in other words, those predicted by general relativity. But what about *real ones* out in space? To see if our theoretical predictions mean anything we need real candidates. Do we have any good candidates? First of all we have to consider where we would find them. We obviously couldn't see a black hole directly; they're far too small. Stars collapse to black holes that are only a few miles across. Our best bet is to look for something that we could observe that would give them away. If a black hole were, for example, moving through a large gas cloud, the gas would be pulled in and would give off X-rays. Even better would be a black hole in a binary or double system. If gas from the star in the system was dragged into the black hole it would become a strong X-ray source. X-rays, therefore, are the key.

In December 1970, the first attempt to find X-rays on a large scale was made. The X-ray satellite UHURU was launched from Kenya in Africa, and indeed it found a large number of sources. Most were soon shown to be neutron stars, but interest finally began to center on a particularly interesting source called CYG X-1. What was needed was an invisible source (black holes cannot be seen) with a mass greater than 3.2 solar masses. Did CYG X-1 qualify?

Let's look at how the X-rays could be generated. Around any binary system there is a double loop that looks like a figure eight where the gravitational potential (field) is the same at all points (see figure 8.10). It is called the Roche lobe. The point where the two lobes join is a critical point called the Lagrangian point. If any gas from the star (we'll call it A) passes the Lagrangian point, it will fall into the black hole (B).

But how could we get gas from A past the Lagrangian point? How would the gas be generated? There are, in fact, two ways. First of all, we know from our study of stellar evolution that A will bloat up toward the end of its life; it will become a red giant, and when it does its outer layers will likely pass the Lagrangian point. It is also possible that if the star is massive enough to begin with (a

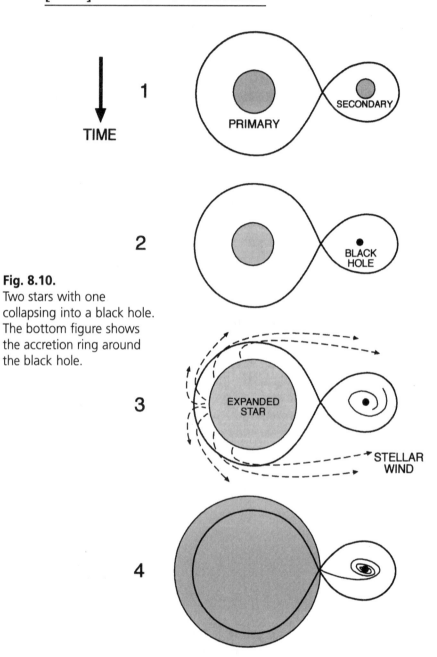

Fig. 8.10.
Two stars with one collapsing into a black hole. The bottom figure shows the accretion ring around the black hole.

blue giant, for example) streams of gas will be ejected from the surface. If any of this gas passes the Lagrangian point it will be pulled into an "accretion ring" around the black hole. From a distance this accretion ring would look like a whirlpool. And as the gas within it whirls closer and closer to the black hole it will be compressed and heated. Furthermore, there will be friction between the various layers, and as a result the gas will be heated to *billions of degrees*. When gas becomes this hot it gives off X-rays.

Was the dark object in Cyg X-1 a black hole? The critical question was its mass; it had to be greater than 3.2 solar masses. As information about the system poured in, astronomers began to understand it much better. The two objects were rotating around one another in 5.6 days and the X-rays changed rapidly at times, indicating it was a very small source. The first thing that had to be determined was the mass of the visible star. First estimates came in at 23 solar masses. With this they could determine the mass of the invisible object. It appeared to have a mass of 10 solar masses, easily enough to be a black hole. In the years since there has been considerable controversy, but Cyg X- I still appears to be our best candidate. It's mass has been determined by independent groups several times since and it has always shown to be considerably greater than 3.2 solar masses.

In 1978 NASA launched the Einstein X-ray satellite and many more X-ray sources were found. Astronomers were sure that some of them would be black holes. They studied them in anticipation, but while they found many good neutron star candidates they found no more black hole candidates. All was not lost, however. Other candidates were soon discovered. One called LMC X-3 is in the Large Magellanic Cloud, a nearby galaxy named for the l6th century explorer, Ferdinand Magellan. LMC X-3 was studied in detail by Anne Crowley of Arizona State University and Dave Crampton and John Hutchings of the Dominion Astrophysical Observatory in Canada. They showed it was an excellent candidate, having a mass slightly over 3.2 solar masses.

Two other candidates, CAL 87 and LMC X-3, are also in the Large Magellanic Cloud. More recently, Yale astronomer Charles Bailyn and several colleagues studied seven binary systems in our galaxy that appeared to contain black holes. All were shown to have an object in them that had a mass over 3.2 solar masses. And another source called GRS 1915+105 was discovered in 1990 using MERLIN, a large radio telescope in England.

So far I've only discussed black holes that have arisen in the collapse of a massive star. But in the next section we will see that there is another type of black hole, some of which may be gigantic. Indeed, they may reside in the cores of galaxies and quasars. One of the best candidates of this type is M-87 in the constellation Virgo. It is an active galaxy that has a spiral-shaped disc of hot gas being emitted from its core. Recently we have been able to "weigh" the object at the center of the galaxy that is causing all the commotion. There seems to be no doubt that it is a black hole. Indeed, many other candidates of this type now exist.[14]

Do We Live in a Black Hole?

This might seem like a crazy question to you, but scientist have been discussing the possibility for years. The first and most obvious problem seems to be that the density of the universe is far from great enough. This is not true. It's easy to show that a black hole with a mass of billions of solar masses would have a density hundreds of times less than that of water. We do, however, need a mass that produces an escape velocity greater than that of light. As it turns out, the overall mass of the universe out to 15 billion light years is close to the amount needed. We're not certain whether it's enough, but it's close. Our universe would also, of course, have to be closed, and we're not sure about that yet either.

Another problem if we are inside a black hole is: Why is the mass not inside a singularity? It is possible that it is still dispersed and will one day collapse into the singularity.

At the present time scientist have mixed opinions on the question, so we're not sure if we are, indeed, living in a black hole, but it is an interesting question.

MINI AND EXPLODING BLACK HOLES

Everything we have said so far has been directed at stellar collapse black holes, those formed in the collapse of a large star. But as I mentioned above, there is another type. We will see in the next chapter that the universe began as an explosion about 15 billion years ago. We refer to it as the Big Bang. One of the first questions we are likely to ask about this explosion is: Was it a uniform, or homogeneous, explosion? In other words, did everything expand out uniformly so that the density throughout the gas cloud remained constant? It's pretty unlikely that this was the case. In fact, we have proof that it wasn't. All we have to do is look around us. If it did expand uniformly, galaxies wouldn't exist; in fact we wouldn't exist either. Galaxies and so on had to form from "inhomogenieties" in the gas cloud. But if inhomogeneities formed, some of the mass of the gas cloud was likely compressed so hard that black holes formed. If so, a whole array of black holes would arise, from tiny, atomic-sized ones all the way up to massive ones that now would be in the cores of galaxies. We refer to them as primordial black holes.

In many ways, the most interesting of these black holes are the tiny ones—the "mini" black holes. They wouldn't be much larger than atoms. If one struck the earth it would pass right through it. Our sun, on the other hand, has enough mass that it would be able to stop one; if so, it would end up in the core.

To see why mini black holes are particularly interesting, we have to go back to the year 1972. Jacob Bekenstein of Princeton University was examining the properties of black holes when he discovered that they could have a surface temperature greater than zero degrees Kelvin (this is the lowest possible temperature in the universe). Prior to this it was assumed that they would have a temperature of zero. After all, they absorb everything and radiate nothing. How could it be otherwise?

To Bekenstein the discovery made no sense and he wasn't sure if he should take it seriously. He worried about it, and although he published it, he didn't follow up on it. In England, however, when Stephen Hawking heard of the result he became particularly intrigued. He convinced himself that black holes would, indeed, have a surface temperature greater than zero. Then, applying quantum mechanics, he saw how it was possible.

Born in 1942, Hawking decided early on that he wanted to be a physicist. He completed his undergraduate work at Oxford, then went to Cambridge for his doctorate.[15] It was during this time that he developed amyotrophic lateral sclerosis, a devastating neurological disease. Within a short time he was restricted to a wheelchair. Today he can no longer speak but uses an electronic voice machine to communicate. Despite his disability he has made tremendous contributions to physics. He soon realized that the key to surface temperatures higher than zero in black holes was what we call virtual particles. They are particle pairs, such as electron-positrons and proton-antiprotons, that pop into existence spontaneously out of the vacuum. Under most circumstances they form and annihilate one another so quickly that we can't observe them, but when these pairs are created in the vicinity of the event horizon of a black hole, the strong tidal forces that are present tend to tear them apart. One of the pairs, for example, may fall through the event horizon while the other escapes to space. To us, at a distance from the black hole, it would appear as if the black hole was emitting particles and radiation.

But if radiation is being emitted, the object has to have a temperature greater than zero (all objects above zero radiate). Bekenstein had been right. Hawking calculated the surface temperatures of black holes of various masses. For a stellar-collapse black hole it was so close to zero that we would never notice the difference. But smaller black holes, particularly mini black holes, had a surface temperature that was significant. Hawking pointed out, in fact, that as a black hole radiated, it lost energy and got smaller, and as it got smaller it gave off more and more energy. Finally, in its last seconds, the energy released would be so great it would appear to us to explode.[16]

What would be left after the explosion? Again, the strange object we called a naked singularity. Shortly after Hawking reported the possibility of exploding black holes astronomers discovered mysterious explosions in space. There was considerable speculation that they might be black hole explosions, but it was eventually shown that they were the result of neutron stars. So far we still haven't detected black hole explosions.

JOURNEY INTO A BLACK HOLE

What would it be like to go into the wormhole of a black hole? In particular, could we use them as subways through space that would allow us to travel long distances in short times? I mentioned earlier that this might be possible in the case of a Kerr black hole.

Several computer simulations of what it would be like to travel into a black hole have been made. C. T. Cunningham of Caltech made some of the first in 1975.[17] He dealt only with Schwarzschild black holes. William Metzenthen of Monash University in Australia extended this to charged black holes in 1990.[18] He showed that as you approached a black hole, you would appear to be entering a long, dark tunnel. A series of rings would appear in the distance. More lighted rings would form as you passed through the event horizon, and as you continued closer to the singularity they would increase in size until they finally merged.

As fascinating as these computer simulations are we still have to ask ourselves: Is it really possible? Astronomers knew there were tremendous difficulties. The first problem is called "pinch-off." Wormholes pulsate, and it's virtually impossible to get through the pulsations (see figure 8.11). The wormhole squeezes down to zero radius as you try to get through. A second problem is that wormholes are one-way. If we went through one to a distant point in space we couldn't get back through the same wormhole. A third problem is the tidal forces that are associated with black holes. We saw earlier that because of the difference in gravitational pull as we approach a black hole, we would be pulled apart. In effect, we would end up looking like a piece of string before we got into the black hole. A fourth problem is radiation. According to Hawking's prediction there would be a tremendous amount of radiation near the singularity and we would, of course, pass it in our trip through the wormhole.

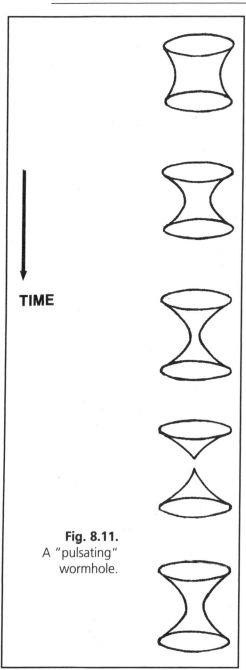

TIME

Fig. 8.11.
A "pulsating"
wormhole.

Last but not least is the problem of an exit. Black holes only pull things in. How do we get out? Einstein's equations show that exits should exist. Scientists refer to them as "white holes." A white hole would be a region where matter is ejected. Do we have any evidence for such objects? Quasars and certain types of active galaxies do appear to be ejecting material from their cores, but it now seems unlikely they are white holes. In fact, Doug Eardley of Yale has shown that if white holes formed in the early universe they would not have survived. This leaves us with a problem—no exits.

All in all, in fact, there appeared to be so many problems with wormholes in the 1960s and 1970s that few scientist believed they would ever become traversible. Wormholes as subways to distant parts of the universe was just science fiction. Then in the mid-1980s things changed.

TIME TRAVEL VIA WORMHOLES

The solution was found almost by accident. Carl Sagan of Cor-

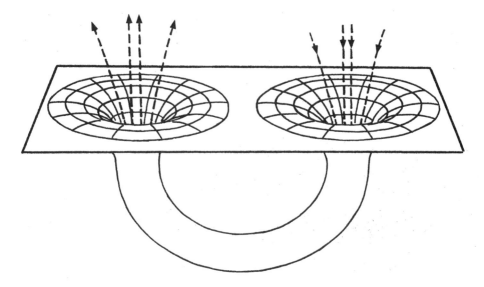

Fig. 8.12. A wormhole in space.
Matter is pulled in at one end and exits at the other.

nell University had written a science fiction novel called *Contact*. It was later made into a movie. In the novel, the heroine, Ellie Arroway, detected a strange radio signal from a distant star. She was able to break the code and translate the signal. It gave the instructions for a "machine" that would allow her to travel through space to the stars.

Sagan assumed that the "machine" was the wormhole of a black hole. Worried about the scientific accuracy of his idea, he sent the manuscript to one of the leading black hole experts, Kip Thorne, who is at Caltech. Thorne read it and was disturbed when he saw that Sagan had used a black hole wormhole. He knew there were too many difficulties for that to be possible, but he wanted to help Sagan. He therefore took a close look at Einstein's field equations to see if there was a way around the problem, and to his surprise he found one.[19]

To be traversible, wormholes had to have several properties. They had to have small tidal forces, and they had to be two-way so that somebody could return through them. Transit times also had to be reasonable. Furthermore, radiation effects had to be minimal and you had to be able to construct the wormhole out of reasonable materials in a reasonable time. And finally there could be no pinch-off.

Thorne found that all of these problems could be overcome. He left many of the details for his student Michael Morris to work out. The greatest difficulty was the pinch-off effect. To overcome it the wormhole had to be lined with special material that could withstand tremendous pressures. We now refer to this material as "exotic matter." It would have to have some very exotic properties, but it is possible that in the distant future it could be constructed.

Within a short time it was shown that these wormholes could be transformed into time machines. In other words, they could be used to take us to the future and past. Of course, if they did allow us to visit the past there was still the problem of causality that we talked about earlier. It hasn't been overcome but there are ways around it.[20]

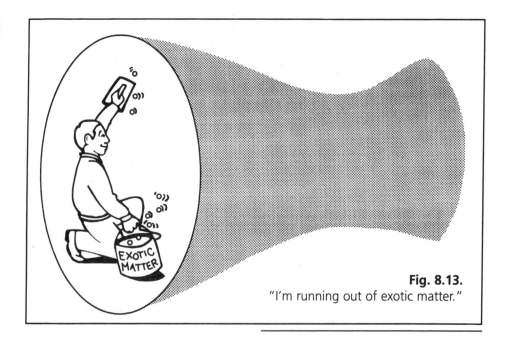

Fig. 8.13.
"I'm running out of exotic matter."

BACK TO EINSTEIN

By the time most of the properties of black holes had been unveiled, Einstein had been dead for a number of years. He died in 1955. Even in his last years he had little use for the idea of black holes, but despite this he was the one most responsible for what we now know of them through his theory of relativity. I should mention, though, that black holes are also predicted in the rival theories. Brans-Dicke theory, for example, predicts them, and as we saw earlier even Newton's theory predicts something that resembles a black hole.

Einstein eventually received the Nobel Prize for his work, but it was not given for his general theory of relativity, or strangely, not even for his special theory of relativity. General relativity was still too controversial when it was awarded to him in 1922. Surprisingly, Einstein was nominated for the prize eight times before it was actually awarded to him. Each time his nomination was rejected.[21] Considering that by this time he was acknowledged by most to be one of the greatest scientists that ever lived this is strange. Why did it happen? One of the main reasons was the German scientist Phillip Lenard. Early on he was a strong supporter of Einstein but when Hitler came to power he became a strong Nazi supporter. He spent much of his time trying to discredit Einstein, and, unfortunately, he had influence with the Swedish authorities that awarded the prize.

Einstein's theories themselves, however, were also a problem. They were simply beyond the comprehension of most of the committee members. Many of the members wanted to give him the award for his special theory of relativity, but they worried that it would eventually be shown to be incorrect. Finally in 1922 they came to an agreement and awarded him the prize for his work on the photoelectric effect.[22]

To the Ends of the Universe

G ENERAL RELATIVITY WAS A POWERFUL NEW tool that would enable Einstein to look at the universe in a new way. Scientists had pondered the mysteries of the universe for centuries, but most of their ideas were speculative. The only mathematical model had been put forward by Newton, but it left many questions unanswered. One of the major problems was the boundary. Did the universe, in fact, have a boundary, or did it go on forever? Even here there was a catch. If it had a boundary, someone would always ask: What's on the other side? On the other hand, if you assumed that it went on forever, you had to define what that meant. An infinite universe was difficult for most people to comprehend.

Newton had faced these problems and Einstein

knew that he would have to face them, too. It wasn't going to be easy. Furthermore, there were many things about the universe that astronomers did not know. We take "galaxies" for granted today. They are the "island universes" of stars that make up the universe.[1] At that time, however, astronomers did not know that they existed. They knew that we lived in a huge system of stars, called the Milky Way, but there were uncertainties about what was beyond it. Einstein assumed that the universe was made up of stars, distributed approximately uniformly. As basic postulates, he assumed that the universe was homogeneous (the same everywhere) and isotropic (the same in all directions).

Using these assumptions he was able to devise a cosmology based on his general theory of relativity. It gave him, in effect, a mathematical model of the universe. But when he went to solve his equations he found his model was unstable; it would either expand or contract. This was unsatisfactory; as far as Einstein knew the overall universe was static. At least that's what astronomers had told him.

Einstein looked over his equations carefully to see how he could overcome the problem. The only way, it seemed, was to add an extra term to them. He was very reluctant to do this. To him, adding a term destroyed much of the "beauty" of the equations. One of the major prerequisites of a theory, as far as he was concerned, was simplicity. The equations had to be elegant and concise, but capable of conveying a large amount of information. An extra term destroyed this to some degree. He finally convinced himself, however, that since the term would affect things only on a very large scale—the scale of the universe—and would have no effect on things as small as the solar system, it would not be problematic. He called it the cosmological constant.

With the added term he was able to solve his equations. The universe he got was stable. It was spherical, like the surface of a ball. Since it was in four dimensions, it was better described as a cylinder in four-dimensions, with the time axis along the axis of the cylinder. The problem of a boundary appeared to be solved. The curvature of the universe came from the matter (the stars) in it, and like the surface of a ball it wound back on itself and had no boundary. If you set out in Einstein's universe, trying to reach the boundary, you would trace out a large circle and end up where you started. How far you would travel depended on the average density of matter in the universe. Einstein heard that Edwin Hubble of Mt. Wilson Observatory in California was working on this problem. He contacted him and got an approximate value for the average density. Substituting it into his equations, he found the universe would have a radius of about ten million light years.

About this time another model of the universe was published. Willem de Sitter of Holland noticed that Einstein had missed a solution. He worked out the details and published it. It was a strange solution that gave an empty universe; in other words, it was a universe with no matter in it. A model universe that contained no matter didn't seem like much of a model. De Sitter admitted that it had problems, but it was known that the average density of the universe was extremely small, and to a first approximation it was empty. The model, therefore, had to be taken seriously. There was, however, another problem. Clocks at a distance from an observer would appear to run slow, which meant that light from distant objects in the universe (assuming there were objects) would be redshifted. This didn't appear to make sense.

Einstein had little use for de Sitter's model. To him an empty universe made little sense, and was hardly worth considering. Eddington also preferred Einstein's universe, but he admitted that de Sitter's universe had some interesting properties.

MORE UNIVERSES

For many years there were two models of the universe: Einstein's and de Sitter's. And no one was sure which one was correct. Things seemed to be at an impasse. Eddington wrote an article on the two theories and published it in *Observatory*. In the article he mentioned that it was strange that only two models had been formulated. It seemed that there should be more. Within a short time he got a letter from Georges Lemaître, a Belgian who at one time had worked for him as a student. Lemaître reminded him of a theory he had shown him many years earlier. At the time Eddington had paid little attention to it, and had completely forgotten about it. Lemaître had published it in an obscure journal. Eddington apologized and asked Lemaître to send him a copy. Lemaître did, and Eddington had it published in *Monthly Notices*, where it would get more attention.

Lemaître's model was a combination of Einstein's and de Sitter's. It started out as a singularity, expanded to an Einstein universe where it stagnated. Finally it began to expand again and ended up as a de Sitter universe.

FRIEDMANN

There were now three models of the universe. Actually, there was a fourth that had been forgotten. It had been put forward many years earlier by Alexandr Friedmann of Russia. Born in 1880, Friedmann graduated from St. Petersburg University in 1910. His main interest was meteorology, but he had a strong mathematical background and when he read Einstein's theory it intrigued him. Einstein had added a term, the cosmological constant, to his equations to keep the universe stable, but he had not developed the theory without the cosmological constant. Lemaître had also used the constant in his theory. Friedmann decided to discard the constant and see what he got. As Einstein did, he ended up with an unstable universe. Following up on the details, he found that there were three possibilities within his theory, and they depended on a particular value of the average density of matter in the universe. We now refer to it as the critical density. If the density of matter in the universe was greater than the critical density, the universe would eventually stop expanding and collapse back on itself. This is referred to as a closed universe; space-time in it is positively curved. (A two-dimensional example of positive curvature is the surface of a ball.) If, on the other hand, the average density of the universe was less than the critical density it would expand forever. In this case the space-time is negatively curved (in two dimensions, like the surface of a saddle) and the universe is open. The dividing line between these two cases is a flat universe; it would also expand forever and is open.

Friedmann sent his paper to Einstein, but received no reply, so after several months he published it. Upon reading the paper Einstein thought he detected an error and sent a note to the editor. The note was published and Friedmann saw it; looking at the criticism he saw that it wasn't valid. Einstein was wrong. Rather timidly he informed Einstein of the mistake and Einstein published a retraction. Einstein was not impressed with Freidmann's theory, however; he viewed it as little more than a mathematical exercise. He was sure the universe was static, and Friedmann's model applied only to an expanding universe. Within a few years, however, it would be shown that the universe was, indeed, expanding. By this time, however, Friedmann was dead; he died of pneumonia in 1925 at the age of 37.

Although it was published in a prestigious journal, there was little interest in Friedmann's model. After Lemaître's expanding universe model became well-known, however, scientists dug it out and took a second look at it. But both Lemaître's and Friedmann's theory created a problem. Both theories assumed that the universe began as a singularity. In other words, there had to be a "beginning" to the universe. Eddington didn't like this. What was here before the universe came into existence? And how did it come into being? These were questions that bothered him. Reluctantly, though, over the years he began to accept the idea.

Lemaître was more comfortable with the idea of a "beginning." He visualized it as a "primordial atom," (a few light years across) and was soon looking into the details. His primordial atom was a nucleus of neutrons. It was well-known at this time that the neutron was unstable and would decay to a proton and an electron (there's actually another particle involved but we won't worry about it here) in about ten minutes if pulled out of the atom. Seizing on this, Lemaître assumed his primordial atom would eventually start to break down. It would spontaneously decay, and out of the byproducts would come the universe we see today. He published his theory in 1931.[2]

THE EXPANDING UNIVERSE

So far we've been discussing models of the universe—theoretical entities predicted by Einstein's theory. But how did they relate to the real universe? Very little was known about the overall structure of the universe when Einstein was formulating his theory. But important advances were soon to come. A well-to-do Boston businessman, Percival Lowell, became fascinated by Mars in the early 1900s and decided to build an observatory to study it. After traveling across America in search of the best site for his observatory, he settled on Flagstaff, Arizona. It appeared to have the best "seeing" in America. Lowell had little interest in the stars, but there had been speculation that the "fuzzy white nebulae" might be solar systems in formation, so he had some interest in them.[3] Equipping his telescope with the best spectroscope money could buy, Lowell was ready to study Mars. A spectroscope is an instrument that splits light up into its various frequencies; it gives us a spectrum of lines that tell us a considerable amount about the object emitting the light.

Lowell hired Vesto Slipher to get the spectroscope set up and working. Slipher had just graduated with his doctorate from Indiana University at Bloomington and had done his thesis on the spectrum of Mars. Slipher arrived at Flagstaff in 1901. For the first few months he spent most of his time getting the spectroscope to work, then over several years he used it to study the spectra of Venus and Mars. The telescope he was using was a 24-inch reflector. Finally in 1909 he was ready to study the white nebulae; he started with the largest one which was in the constellation Andromeda.[4] With some difficulty he finally managed to get its spectrum. To his surprise the lines were shifted from where they should have been. Shifts of this type had been encountered before; they meant that the object was moving either away from you or toward you. If the shift was toward the blue end of the spectrum the object was moving toward you; if it was toward the red end of the spectrum it was moving away from you. In this case the shift was toward the blue end of the spectrum. What was particularly surprising, however, was the magnitude of the shift. It indicated that the Andromeda Nebula was moving toward us at the incredible speed of 300 Km/sec. This was much larger than anything that had ever been encountered. It worried Slipher so he turned his spectroscope to a similar object in the constellation Virgo and took its spectrum. It was dimmer and more difficult to obtain, but finally he got it and this time he was in for even more of a surprise. The object in Virgo was traveling three times faster than the one in Andromeda. Even more amazing, it was red-shifted, indicating that it was moving away from us rather than toward us. Slipher didn't know what to make of this. He informed Lowell, who was now back in Boston, and Lowell told him to continue looking at more of the objects. By 1914 Slipher had the spectra of fourteen nebulae. Most were red-shifted, but a few were blue-shifted. He presented his results at the American Astronomical Society meeting in Evanston, Illinois. No one knew what to make of his results, but everyone at the meeting was impressed; they realized it was an important discovery. He got a standing ovation.

What was causing the shifts? Slipher was sure it was a result of the motion of our galaxy, the Milky Way. According to him we were moving toward the Andromeda Nebula and away from most of the others. Slipher continued his work and by 1922 he had the spectra of forty-one white nebulae. He was now nearing the limit of his 25-inch telescope. Almost all the objects were red-shifted, but Slipher clung to his view that they were a result of our motion, or more exactly, the motion of our galaxy.

A number of astronomers now began to remember that de Sitter had predicted that if two objects were placed in his universe they would move away from one another. With almost all of Slipper's shifts being red shifts it seemed as if everything was moving away from us. Furthermore, the dimmer the galaxy, the greater the redshift. A number of scientists began to wonder if there was a relationship between recessional velocity (speed away from us) and distance. Speculation abounded and a number of people tried to show that there was a relationship, but nothing could be proved.

Then came Edwin Hubble. He had been a graduate student in 1914 and was in the audience when Slipher announced his results.

Hubble's thesis had been on white nebulae, and he was hoping to get a job at Mt. Wilson Observatory when he graduated. But World War I intervened and he was drafted. In 1919, however, he returned to the United States and was soon on his way to Mt. Wilson. On the summit was the largest telescope in the world, the 100-inch Hooker reflector. Hubble planned on continuing his work on white nebulae. One of the first things he would have to determine, however, was what these strange objects were. Lowell thought they were solar systems in formation. Hubble had his doubts.

Were they composed of stars or gas? Hubble took long exposures, studying each carefully, looking for sign of stars. Finally, in the outer regions of several of the objects, he was able to identify individual stars. Looking at them closely he saw that some of them were a type of star known as a cepheid variable. Cepheid variables change periodically in brightness; in essence, they pulse.[5] What was particularly important, though, was that a relationship between their pulsation period, the average brightness of the star, and the distance to it had been established. This meant that if you could measure its period of pulsation and average brightness, you could determine its distance. Using cepheids, Hubble was able to determine the distance to the Andromeda Nebula. It was beyond the edge of our galaxy. Indeed, it was far beyond the estimated limits of our galaxy. Hubble then turned to other white nebulae, and showed that they were also outside the Milky Way. The universe, it seemed, was composed of "island universes of stars"—what we now call galaxies. He soon showed that there were hundreds of these galaxies around us (we now know there are billions).

Hubble was familiar with Slipher's work, and it now made more sense. Furthermore, he was familiar with de Sitter's prediction that the objects in the universe should be moving away from one another. He wondered: Was there a relationship between recessional velocity and distance for galaxies? Hubble wanted to find out, but he knew there had been controversy over the years, so he proceeded cautiously. Using cepheid variables he determined the distance to six of the brightest galaxies. The redshift of each of them was known—obtained by Slipher. Hubble made a plot of recessional velocity versus distance. There was consider-

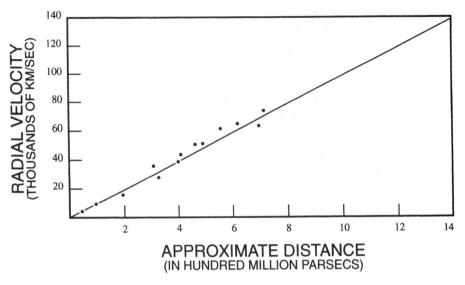

Fig. 9.1. Hubble's plot of radial velocity versus distance.

able scatter in the points, but to a first approximation a straight line could be drawn through them (see figure 9.1). There appeared to be a relationship between the two quantities. Hubble went on to look at more galaxies and in 1929 he published his results. The relationship was tentative, but the evidence was strong. Hubble mentioned de Sitter's prediction in his paper, hoping it would give his conclusion some support.

It was obvious at this point, however, that a lot more work was needed. Hubble knew he would have to probe much farther into the universe and with the help of an assistant, Milton Humason, he went after more distant galaxies. He could see cepheids in the nearest galaxies, but they were not visible in the more distant ones. He therefore turned to other indicators. He used the brightest stars in galaxies, then finally the brightness of the galaxy itself to determine their distances. The method was crude, but it proved effective. By 1931 he and Humason had probed galaxies out to 100 million light years. Very large red shifts were now being obtained, and when he made the plots the results were clear: there was a relationship between recessional velocity and distance, and it was linear. This meant that a galaxy twice as far away was traveling twice as fast. Furthermore, it meant that the universe was expanding, as if it had begun in an explosion. Hubble was cautious about saying anything about an expanding universe. But by 1936 when he published his classic "Realm of the Nebulae," he was convinced that it was, indeed, expanding.[6]

THE LAUGHING GIANT

If the universe was expanding, it had to have a beginning. One way to convince yourself of this is to imagine taking a movie of the expansion, then running the movie backward. Eventually all the galaxies will get back to a point. As we saw earlier, Lemaître had considered this point; he referred to it as the primordial atom and assumed that it gave rise to the universe, but he thought in terms of a "breaking down" of the original nucleus.

George Gamow became interested in the problem shortly after World War II. He had been working on the atomic bomb and as a result soon began thinking of a "building up" of the elements, rather than a breaking down. Born in Odessa, Russia, in 1904, Gamow received his Ph.D. in the late 1920s. This was the heyday of quantum mechanics and he was soon caught up in the excitement. A larger-than-life man with a tremendous sense of humor, he loved to play practical jokes. For him, physics was fun.[7]

Intrigued with Lemaître's ideas, Gamow assumed the particles of the universe were built up via "fusion." The energy and temperature of the early universe was incredibly high. Particles therefore slammed into one another with a tremendous force and many fused. Starting with a dense gaseous nucleus consisting mostly of neutrons, Gamow visualized a process in which all the elements were built up, one after the other. A proton, for example, would hit a neutron and become a deuterium nuclei, or deuteron. Then the deuteron would be hit and become a heavier nuclei. His graduate student, Ralph Alpher, worked out the details for his thesis. The theory eventually became known as the Alpha-Beta-Gamma theory (the first three letters of the Greek alphabet). According to it, all the elements of the universe were produced in a fusion process. It was an ingenious idea and Gamow and his student received considerable publicity over it.

But not everyone was convinced they were right. Enrico Fermi of the University of Chicago began looking at the process, and he saw problems at elements 5 and 8. There appeared to be gaps at these nuclei that could not be overcome. About the same time Gamow began to notice the same problem and it soon became obvious that the seemingly ingenious solution to the problem of where the elements of the universe came from wouldn't work. A few of the light elements such as helium would be produced, but nothing beyond them.[8]

If the elements were not produced in the explosion that created the universe, where were they produced? It seemed that there was only one other possibility: the cores of stars. They were the only regions where temperatures were high enough, and in 1956 Fred Hoyle, Margaret and Geoffrey Burbage, and William Fowler showed that this, indeed, was the case. The elements up to iron were produced in the cores of stars, and as the stars exploded heavier elements were produced. They were distributed to space in the explosion.

A COMPETITOR

The theory that the universe began as an explosion is now known as the Big Bang theory. It is perhaps ironic that the person who gave the theory it name, Fred Hoyle, was also one of the founders of the Big Bang theory's first competitor. Hoyle became interested in cosmology and the Big Bang theory in the mid-1940s; he got together with Hermann Bondi, an expert in fluid dynamics, and a graduate student, Thomas Gold, to learn more about it. After studying the theory for a while, they began to get skeptical of parts of it. They attended a movie one evening that had no ending (the beginning of the movie was the same as the end). As they came out of the theater one of the group suggested that maybe the universe was like the movie they had just seen. In other words, maybe it had no ending or no beginning, and was always the same. It would be a "steady state" universe if this was the case. To have such a universe, however, they would have to have matter pop into it to compensate for the expansion. The amount would be small, but it would violate the principle of conservation of matter. They discussed the problem and decided not to worry about it. After all, the Big Bang theory also violated conservation; the only difference was that the mass all came at once in the Big Bang theory, while it evolved slowly in the Steady State theory. The group worked out the details, then after a small disagreement, they decided to publish separately. Hoyle would publish one paper, and Bondi and Gold the other.

Soon the new theory was being taken seriously. There were now two theories: the Big Bang theory and the Steady State theory. Which one was correct? Controversy continued throughout the 1950s and into the 1960s. Each theory had several well-known supporters. But as more and more evidence came in, most of it appeared to go against the Steady State theory. Then in 1965 came the final blow: the discovery of background radiation in the universe.

RADIATION FROM EVERYWHERE

In the early 1960s Robert Dicke of Princeton University began taking an interest in cosmology and began lecturing on it at his Friday night seminar. He soon found that he preferred the Big Bang theory to the Steady State theory, but the Big Bang theory still seemed to have problems. What, for example, was here before the Big Bang? Dicke decided that the most satisfactory answer was another universe. He pictured a universe that oscillated, with each cycle beginning with a Big Bang, and ending with a collapse. But if this was the case, there should be something in our universe that was left over from the previous universe. In particular, there should be some remnant radiation. Dicke assigned the details to a junior colleague by the name of Jim Peebles. Peebles was to calculate the temperature of this remnant radiation, and determine any other properties it might have. About the same time Dicke asked

Peter Roll and Dave Wilkinson to set up a small detection device to see if they could detect it.

Peebles made the calculation and came up with a temperature of approximately 10° K. Roll and Wilkinson began building their detecting device, but before they could complete it, fate intervened. Two scientists at Bell Labs in Holmdel, N.J., found the radiation.

Arno Penzias and Robert Wilson were working on the Telstar project for Bell Labs. Telstar was a communication satellite that had just been launched. To communicate with it, Bell technicians had built a large, strange-looking radio telescope that resembled a giant horn. What was particularly exciting to the two men was that they could use it for their own astronomical research after the Telstar project ended. Penzias and Wilson couldn't wait. Finally in 1964 they were able to start their research project. They began by testing the telescope for noise; they would have to get rid of all extraneous noise before it could be used for astronomical purposes. To their annoyance they soon discovered a persistent hissing. They tried everything to get rid of it, including taping all seams on the horn, but regardless of what they did, some of it remained. Their research project appeared to be doomed.

Then one day radio astronomer Bernard Burke was attending a lecture by Jim Peebles. Peebles talked about their work and the radiation they believed pervaded all space. He said it should be detectable and announced that several people at Princeton were setting up a radio telescope to look for it. Shortly after the lecture Burke received a telephone call from Penzias. Out of courtesy Burke asked him how his project was going. Penzias told him about the problems they were having. Peebles's talk came to mind and Burke told Penzias to phone Dicke. Dicke might have the answer to their problem, he said. And, as they say, the rest was history. Penzias phoned Dicke, and Dicke and several colleagues visited Penzias and Wilson at Holmdel. Soon there was no doubt: the annoying hiss was the cosmic background radiation that Dicke and his group had hoped to find. The two groups published separate papers in 1965.[9] Dicke's group concentrated on the theoretical prediction of the radiation. Penzias and Wilson concentrated on the radio telescope results, and the detection of the radiation. Penzias and Wilson were eventually awarded the Nobel Prize for their work. Dicke and his group did not share in it, mostly because Gamow and his students had predicted the radiation several years earlier. They were considerably off, however, in their estimate of its temperature.

Peebles initial calculations had also been off. The radiation turned out to have a temperature of approximately 3° K.[10] Peeble showed later, however, that the correct prediction was, indeed, very close to 3° K. When the initial excitement had died down, everyone realized that much more was needed to prove beyond a doubt that the radiation was, indeed, from the Big Bang. Only one point from the radiation spectral curve had been obtained. For proof, the intensity of radiation over a range of frequencies, namely, a plot of the intensity of the radiation versus frequency for 3° K (called a blackbody curve), was needed. Roll and Wilkinson soon got another point on the spectrum. Then as others joined in the search, many other points were obtained.

But there was a problem. The curve had a peak, and all points that had been obtained were on one side of the peak. The other side, unfortunately, was cut off by our atmosphere. To get points in this region, scientists had to get above the lower layers of our atmosphere, which meant that they had to use rockets or balloons. In one of the first attempts to get points in this region a rocket was used. But much to the dismay of the scientists, the points on the other side of the peak did not appear to be on the curve. To their relief, however, they soon found that the exhaust of the rocket had thrown off the data. Finally, however, D. P. Woody and Paul Richards of the University of California obtained a curve that was close to the expected 3° K curve. In 1987 Paul Richards obtained new data, and again the curve turned over where it was suppose to, but there appeared to be excess radiation at several points on the spectrum.

The problem was finally solved when the satellite Cosmic Background Explorer (COBE) was launched in 1989. It had highly sensitive equipment aboard and within a few minutes of launch it verified the 3° K blackbody curve to an accuracy of 1 percent (see figure 9.2). There was no doubt now: it was the cosmic background radiation.

One of the first questions in relation to the radiation after it was discovered was: Is it isotropic? In other words, is it the same in all directions? Penzias and Wilson showed that to an accuracy of about ten percent, it was isotropic. Still, it was quite possible that there was an anisotropy (different in different directions) on a smaller scale. More sensitive measurements would have to be made, and when they came there was no doubt: The radiation was anisotropic. Using a NASA U-2 spy plane, Richard Muller, George Smoot, and Marc Gerenstein of the University of California showed that the temperature of the radiation was maximum in the direction of the constellation Leo, and minimum in the opposite direction. This meant that we were moving through the radiation with a velocity of 600 Km/sec in the direction of Leo. In fact, it was later shown that not only our galaxy but all the nearby galaxies were also traveling in this direction with the same speed.[11]

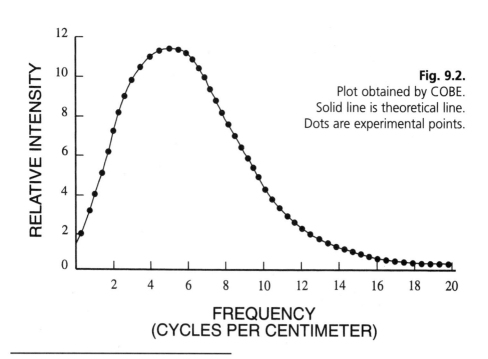

Fig. 9.2.
Plot obtained by COBE.
Solid line is theoretical line.
Dots are experimental points.

FREQUENCY
(CYCLES PER CENTIMETER)

CREATION

In accepting the Big Bang Theory, astronomers had assumed that the universe was created in a gigantic explosion about 15 billion years ago. It's difficult, if not impossible, to imagine what this explosion would have been like. Nevertheless, there is a scientifically correct way of looking at it.

First of all, the explosion did not occur in infinite space as you might think. It actually created space, and all the matter in the space. Secondly, if we look out into the universe today we see all the galaxies expanding away from us. Does this mean that we are at the center of the universe? No, the expansion would appear the same regardless of where we were in the universe. The galaxies recede from everyone in the same way. It might seem that this is impossible, but it isn't. If the galaxies all move away from one another, that is, if the space between them expands, all galaxies will appear to be moving away from everybody. The best way to visualize this is to think of the galaxies as spots glued to a balloon. When the balloon is blown up they separate, and the more you blow it up, the more they separate.

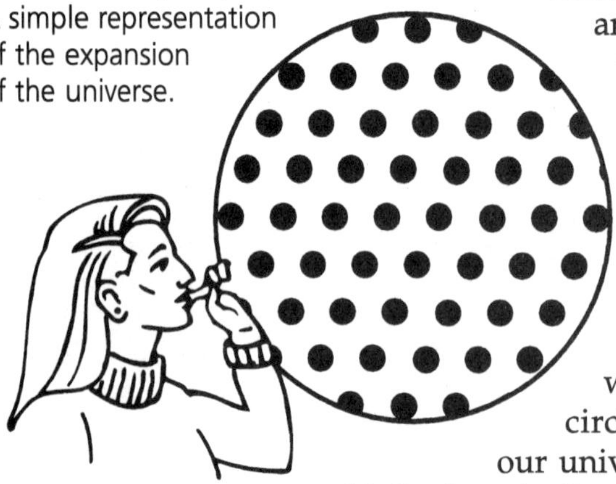

Fig. 9.3. Girl blowing up a balloon with spots pasted on it. A simple representation of the expansion of the universe.

Another question you might ask is: Does the universe have an edge? Again, using the balloon, we can easily see that it doesn't. If we set out on its surface to find an edge we would just go around in a circle. From Earth, however, our universe does have an "observable" edge. As I mentioned earlier, if we look outward from the earth we see galaxies moving away from us at ever increasing speeds. If we look twice as far the galaxy will be moving twice as fast. Eventually, if we look far enough the galaxies will be traveling near the speed of light, and

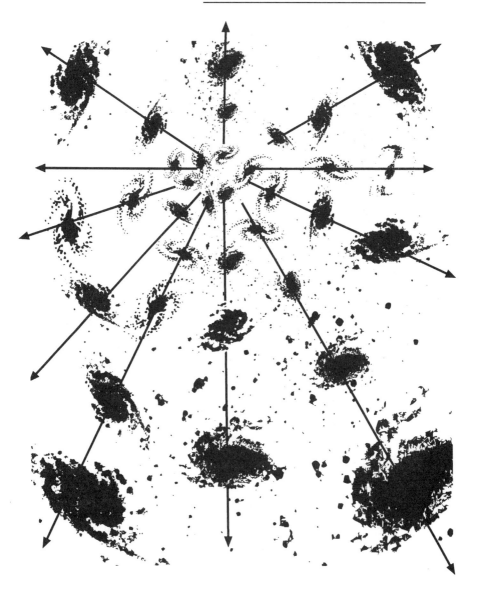

Fig. 9.4. The expanding universe. Galaxies are moving away from one another.

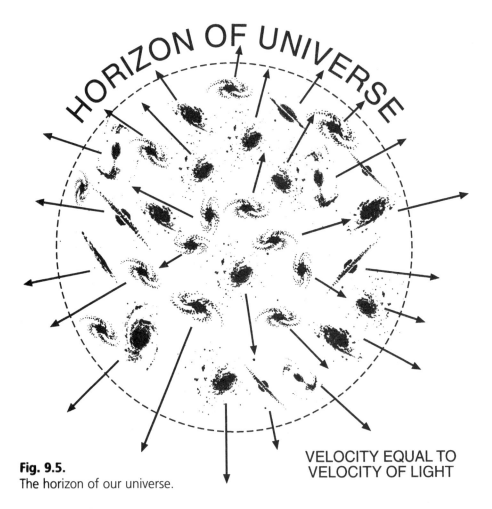

Fig. 9.5.
The horizon of our universe.

VELOCITY EQUAL TO
VELOCITY OF LIGHT

just beyond this they are, in theory, moving at a speed greater than that of light. But we know this is impossible. We therefore see nothing beyond this point.

Another way of looking at this is to consider the time light takes to get to us. If a galaxy is 100 light years away, it takes 100 years for its light to get to us. Similarly, if it is one billion light years away, it takes one billion years for its light to get to us. But we know that the universe is only 15 billion years old, so if we look beyond 15 billions light years there will be no universe. It didn't exist at that time.

It might seem a little crazy, but there actually is more universe beyond 15 billion light years, the horizon of our galaxy. To see why, we merely have to consider two observers, one on Earth and another on a galaxy near the edge of our observable universe. Remembering that everyone can see out to a particular radius and that they are at the center of their circle, we can easily see that an observer near the edge of our universe will have a different universe around him. In particular, his universe will include galaxies that are not in our observable universe (see figure 9.6).

I mentioned that our universe is 15 billion years old. Over the years there has been considerable controversy about its age, however. The first and perhaps the simplest method of determining it comes from the plot that Hubble made. We saw that he plotted recessional velocity versus distance to galaxies. This is now known as a Hubble plot, and the slope of the line that he got in the plot is called H in his honor. To a first approximation the age of the universe is just $1/H$. It was obvious early on, however, that this was only a rough approximation. Hubble's first plot, in fact, gave the embarrassingly low age of 2 billion years. Geologists had found rocks older than this, and Hubble's estimate was therefore suspect.

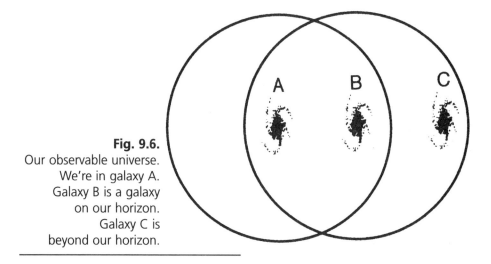

Fig. 9.6.
Our observable universe.
We're in galaxy A.
Galaxy B is a galaxy
on our horizon.
Galaxy C is
beyond our horizon.

Hubble soon found an error, however, and revised his estimate, making the universe much older.

In 1953 Hubble died, but his work was carried on by Allan Sandage, a young astronomer who had just received his Ph.D. Sandage was worried about the responsibility that had suddenly been thrust on him, but he forged ahead with Hubble's program. Hubble had used a "cosmic ladder" to push farther and farther into space. The first rung of his ladder was cepheid variables, the second rung the brightest stars in a galaxy, and so on. As it turned out, Hubble had made mistakes, and his ladder had to be put on a much firmer foundation. One question of importance was: Did the galaxies eventually begin to slow down, or decelerate? Sandage attacked the problem with determination. He worked on it through the 1950s and 1960s, pushing outward to dimmer and dimmer galaxies. Of particular interest to him was whether the straight line in Hubble's plot remains straight, or whether it eventually begin to curve. This would tell us if the outermost galaxies were, indeed, slowing down. Sandage eventually showed that it did curve, and according to his results it curved enough to eventually stop the expansion of the universe. His calculations indicated that it would continue to expand for another 40 billion years, then it would begin collapsing. They also showed that the age of the universe was about 20 billion years.

Sandage was the authority and for years he went unchallenged. Whatever he said was considered to be gospel. Then in the 1970s things began to change. A team consisting of Richard Gott, James Gunn, David Shramm, and Beatrice Tinsley published a paper showing that there was strong evidence that the universe was open.[12] They were sure Sandage was wrong. Another challenge came from Gerard de Vaucouleurs of the University of Texas. De Vaucouleurs had spent years studying the problem and had come up with a different cosmic ladder. He suggested that Sandage had neglected the gravitational pull of nearby clusters of galaxies and that his estimate was in error because of this. De Vaucouleurs came up with an age of 10 billion years.

How could the two estimates be so different? Another estimate was obviously needed, and it was best if it was based on an independent method. It came in 1972. Brent Tully and Richard Fisher were studying the rotational rates of galaxies when they realized that their technique could be used to determine the age of the universe. Many corrections were needed, but Tully and Fisher were finally able to get an estimate. It was in approximate agreement with de Vaucouleurs'. The scientific world was stunned, but many soon began to criticize Tully and Fisher's method. And there was indeed a problem: the obscuring dust and gas between us and the galaxies.

Marc Aaronson of Steward Observatory in Arizona and John Huchra of Harvard heard about the controversy and decided to look into it. They soon saw a way around the problem. Tully and Fisher had used blue-sensitive photographic plates and had made corrections for the dust. Aaronson and Huchra would use infrared plates that would penetrate the dust. No corrections would be needed. They made their first observations in the direction of the constellation Virgo, and to their surprise they got an age between that of de Vaucouleurs, and Sandage. When de Vaucouleurs saw their estimate he encouraged them to take observations in other directions, away from Virgo, and when they did, they got an age very close to de Vaucouleurs, namely, 10 billion years.

Sandage was unconvinced by the announcement. He was now working with Gustav Tammann of Switzerland, and he continued on with his program. His estimate remained at 20 billion years.

Was there a way out of the dilemma? Indeed, there was. There were several other methods of determining the age of the universe. A study of globular clusters (systems of several hundred thousand stars) gave a good estimate and radioactive decay gave another. The radioactive method is based on the speed at which radioactive elements decay. Their decay time is characterized by what is called half-life. By observing the half-lives of elements in the universe it was possible to determine the age of our galaxy, which in turn gave an estimate of the age of the universe. As expected there was a range of values using the two methods. Astronomers decided to average them, and the average turned out to be about 15 billion years. This is now the value that most accept.

Now that we know a little more about the Big Bang theory, let's take a closer look at the details. What was the explosion like? So much work has been done in the last few years we cannot possibly cover it all. I will therefore give only a brief overview. Astronomers prefer to divide the events of the early universe into what are called "eras." Using general relativity, they have managed to trace events "almost" all the way back to the Big Bang. As we go back in time the temperature and density of the universe increase. Near the initial singularity they are extremely high, so high in fact that general relativity breaks down. This is the era of quantum cosmology; to penetrate it we would need a quantum version of general relativity and we don't have one. John Wheeler of Princeton University, however, has assumed that there was a "quantum foam" present at this time. He visualizes it as a region of unimaginable chaos, where particles as we know them did not exist.

At 10^{-43} seconds the universe passed out of this region into what is called the Hadron era.[13] General relativity is now valid, and can tell us something about this era. At the beginning of the era, hadrons, which are heavy particles such as protons and neutrons, didn't exist. Only the particles that make them up, namely, quarks (and antiquarks), were present. At that time they were free, but we no longer see free quarks; in fact, we don't see quarks at all. They are trapped inside the particles they make up. As the temperatures continued to drop, the free quarks suddenly assembled themselves into "bags" and became recognizable particles. This event is called the quark-hadron transition. The universe then consisted mostly of hadrons (and antihadrons).

But hadrons decay to light particles and when the temperature got down to 100 billion degrees, about one ten thousandths of a second after the Big Bang, it entered the Lepton era (leptons are light particles such as electrons). The main particles present now were electrons, antielectrons or positrons, neutrinos and their antiparticles. There was still a few protons and neutrons, left, but they were decaying rapidly. Decaying particles, however, give off radiation, and soon the universe was overwhelmed with radiation. About 20 seconds after the Big Bang the universe entered the Radiation era. The temperature was about 10 billion degrees.

Within the Radiation era an event of particular importance occurred: the appearance of the first nuclei. As we saw earlier,

Gamow suggested that all the elements were produced in the Big Bang. He was partially right; some of the light elements were. When the temperature was down to a billion degrees, about three minutes after the Big Bang, it was cool enough for the first nuclei to begin forming. When a proton hit a neutron, for example, it produced a nucleus of deuterium. The collision of two deuterium nuclei then created a helium nucleus. But as we saw earlier, this is about as far as it got. Small amounts of tritium and lithium were produced, but nothing beyond them.

The universe continued to cool and expand for thousands of years. Little happened during this time, but finally at a temperature of 3000° K electrons and protons began to combine to form hydrogen atoms. As they formed they released radiation (until then, it had been coupled to the matter) and it expanded out into the universe and cooled. This is the cosmic background radiation we talked about earlier. It now has a temperature of 3° K.

The giant cloud of radiation and matter continued to expand, but gradually after about a million years, fluctuations began to form and the cloud started to break up. The first stages of galaxies began to form, and soon stars formed within them. This is referred to as the Galaxy or Matter era.[14]

THE FORCES OF NATURE

In our discussion so far we have left out several important events that are related to the forces of nature. You are no doubt familiar with some of these forces. The gravitational field is the best known; it is the force that holds you to the earth, and holds the earth in orbit around the sun. Another of these forces is the electromagnetic force; it holds the electrons in orbit around the nucleus of the atom. The last two are the ones you may not be familiar with; they are the strong and weak nuclear forces. The strong nuclear force holds the particles of the nucleus—the protons and neutrons—together. The weak nuclear force doesn't hold anything together; nevertheless, it's important in such things as radioactive decay.

These are the forces of nature, and as far as we know, they are the only four. They differ significantly in strength and range. The gravitational field is by far the weakest—10^{39} times as weak as the strong nuclear force. The strong nuclear force is 100 times stronger than the electromagnetic force. The weak nuclear force, on the other hand, is about a thousand times weaker than the electromagnetic force. A further difference is range. Both the gravitational and the electromagnetic force fields have an infinite range. The nuclear force field, on the other hand, acts only over a very short distances—about the size of the nucleus.

Scientists have shown that all four forces may have been together as a single "Primeval force" when the universe began. As the universe expanded, however, they separated. One can think of them as "condensing out." The first to separate was gravity (shortly after 10^{-43} seconds). Then came the strong nuclear force at 10^{-34} seconds and finally the strong and electromagnetic forces separated at 10^{-12} seconds, giving us the four distinguishable forces we see in nature today (see figure 9.7).[15]

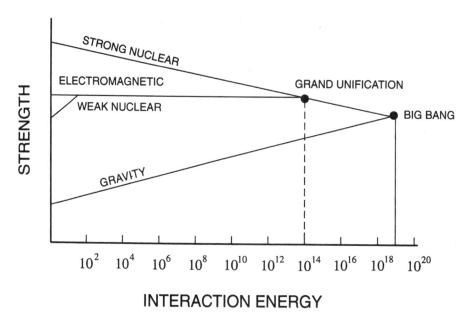

Fig. 9.7. Separation of the four forces of nature as the universe cools.

PROBLEMS OF THE BIG BANG

Despite its tremendous success, the Big Bang theory still has a number of difficulties. The first, called the flatness problem, was brought to our attention by Robert Dicke in 1969. It is best explained by introducing a quantity called omega. It is the ratio of the average density of matter in the universe to the critical density.[16] If omega is greater than 1 the universe is closed; if it is less the universe is open. What is omega for our universe? We're still not sure, but it's somewhere between .1 and 3.

Dicke showed that if omega is so close to 1 now, the universe had to have started out at exactly 1.0 (to 15 decimal places). If it started out slightly less than 1, say, 1.01, it would now be of the order of thousands. This seemed to imply that omega was, indeed, exactly 1 and the universe was flat.

A second problem is called the horizon problem. It was brought to our attention in 1969 by Charles Misner of the University of Maryland. Misner considered the following: suppose you observe a quasar near the edge of the universe, say, at a distance of 10 billion light years. If you turn in the opposite direction and observe another one at a distance of 10 billion light years, they will be separated by 20 billion light years. How is this possible? The universe is only 15 billion years old. Furthermore, if we measure the temperature of the background radiation near the first quasar, it will be 3° K. And if we measured the temperature near the second one it will also be 3° K. How could they be exactly the same? It would take 20 billion years for a message from one to reach the other, and again this is greater than the age of the universe.

A third problem is called the monopole problem. To explain it we have to begin with magnetic monopoles. You no doubt know that a magnet has a north and a south pole. But if you cut the magnet in half, it still has two poles. In fact, if you continue this indefinitely you will still have two poles attached to one another. Magnetic poles don't seem to come "single" as positive and negative charges do. But strangely enough they are predicted to exist separately. Why haven't we seen any monopoles? No one is quite sure. They are predicted to be very massive, and this may be part of the problem. In fact, they are so massive that they could have been created only in the very early universe.

A fourth problem is: How did the galaxies form? The Big Bang theory does not tell us, but we do have some ideas. As I mentioned earlier, density fluctuations (slight variations in density from place to place) likely formed in the original gas cloud as it expanded, and these fluctuations eventually caused it to break up. The details, however, are still fuzzy. We're not sure what caused the fluctuations, or exactly what form they took. Furthermore, there are many different possible paths from fluctuations to our present universe.[17]

A LUMPY UNIVERSE

Soon after Hubble showed that the "white nebulae" were galaxies, Fritz Zwicky of Mt. Wilson Observatory noticed that some of them appeared to be grouped in clusters. In fact, clusters of galaxies appeared to be common. Eventually it was shown that our galaxy, the Milky Way, was part of a cluster, now called the Local Cluster. Then in the early 1950s, Gerard de Vaucouleurs showed that clustering took place on an even larger scale. He found clusters of clusters, what we now refer to as superclusters. Then large voids (regions with no stars) were found between the clusters. The universe was obviously much more complex than we thought.

Jim Peebles of Princeton University decided to make a two dimensional plot of about a million galaxies to see if he could identify a large-scale structure. The results were startling. Filaments, knots, and voids were clearly visible. The universe on a very large scale definitely appeared to be lumpy.

Marc Davis, who was then at Princeton University, and John Huchra of Harvard decided to follow up on Peeble's work. They would do a large-scale redshift survey. This would allow them to make a three-dimensional plot of the galaxies. The two men started their survey in 1978, and by the early 1980s they had completed it. They were excited as they made the plots. The filaments, voids, and so on seen by Peebles weren't an illusion; they were still there in three-dimensions.

Davis left the project and Huchra had to find another partner. He soon teamed up with Margaret Geller of Harvard, and together they extended the earlier work that he and Davis had done. Because of the overwhelming number of galaxies in deep space, they had to limit themselves, so they made deep narrow probes in the form of wedges, like pieces of pie (see figure 9.8). As they completed the first couple of wedges, they noticed large voids with strings of superclusters strung across them. When the fourth wedge was completed in 1989 they noticed the beginning of another structure. A huge sheet of galaxies was visible. It contained tens of thousands of galaxies and was the largest structure ever seen in the universe. They called it the Great Wall.

A different type of survey was initiated in the early 1980s by David Koo and Richard Kron, who were then at the University of California at Berkeley. They took very narrow pencil-like probes and pushed much farther into space than Huchra and Geller had. They saw the Great Wall, but beyond it they saw several other walls. The universe appeared to be made up of great walls. On a very large scale it seemed to have a honeycomb structure.

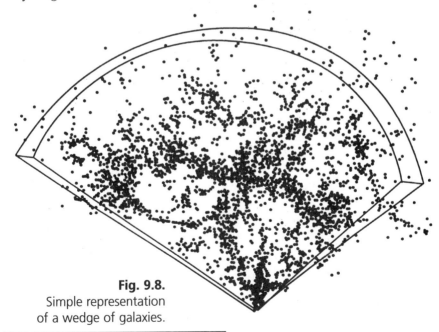

Fig. 9.8.
Simple representation
of a wedge of galaxies.

Redshift surveys are an excellent way of determining the large-scale structure of the universe, but there is another way. As we saw earlier, the universe is made up of clusters of galaxies, superclusters, and huge voids. In other words, the matter is not uniformly distributed, and as a result, galaxies are pulled in directions where there are overdensities, or excesses of galaxies. The velocity that a given galaxy acquires in this way is called its "peculiar" velocity. In addition, of course, there is its velocity from the expansion of the universe; it is called the Hubble velocity.

In recent years peculiar velocities have told us a lot about the universe. Our galaxy and the galaxies around it are part of what is called the Local Supercluster.[18] The largest cluster within it, called the Virgo supercluster, lies at the center. We lie off to one side. The Virgo supercluster was a huge cluster and it was thought to be responsible for most of our peculiar velocity. It was shown, however, that the Virgo Supercluster was not responsible for all of it; something beyond it was attracting us. Astronomers searched for the object and found a huge conglomeration of superclusters now called the Great Attractor. It is believed to have a mass twenty times as great as the Local Supercluster.

INFLATION THEORY

Earlier we talked about some of the difficulties of the Big Bang theory. There's also the problem of the incredible amount of energy that was needed for the generation of the particles in the Big Bang. Where did it come from? An ingenious solution to many of the difficulties was found by Alan Guth of MIT in 1980. He showed that if a sudden inflation occurred in the expanding cloud of the early universe, many of the problems were overcome. In particular, it would explain where the energy came from. According to Guth, this inflation occurred between 10^{-35} seconds and 10^{-33} seconds (see figure 9.9). Guth suggested that the universe settled into a supercooled state just before 10^{-35} seconds, and it created an incredibly fast expansion—much faster than the normal expansion rate of the Big Bang. Between 10^{-35} seconds and 10^{-33} seconds the universe doubled in size a hundred times.

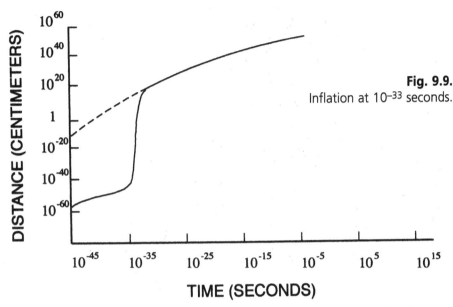

Fig. 9.9.
Inflation at 10⁻³³ seconds.

Examining the consequences of this sudden expansion, Guth found that it overcame the flatness problem, the horizon problem, the monopole problem, and to some degree the galaxy problem. In addition it explained where the energy of the universe came from, namely, it was self-generating. There was, unfortunately, a difficulty in Guth's theory: he couldn't get the inflation to end properly. The problem, however, was solved in 1981 by Paul Steinhardt and Andreas Albrecht of the University of Pennsylvania, and independently by Andrei Linde of Russia. Even with this, however, the theory still had difficulties and it is still not accepted by everyone.

FATE OF THE UNIVERSE

One of the major goals of cosmology is to explain what is eventually going to happen to our universe. We saw earlier how Friedmann's theory tells us that it will either expand forever or stop expanding and collapse back on itself, depending on its average density. If its density is greater than a critical value of about 6×10^{-30} gms/cm³, the expansion will stop and the universe will collapse. If it is less than this it will expand forever. This means that we need an accurate value for the average density. This, unfortunately, has not been easy to get. A careful measurement of the Hubble constant can also tell us if the universe is open or closed.

One way of looking at this problem is to consider the outermost galaxies in the universe. We know that the mutual gravitational attraction of the galaxies will eventually slow them down. Will they slow them down enough to stop the expansion? This, of course, depends on the average density of matter. We can easily make an estimate of the density of visible matter, but it turns out that there is considerable matter we cannot see, and that is the problem. The visible matter is considerably short—about 100 times—of the amount needed to close the universe.

Our problem, then, is the matter we cannot see. We sometimes refer to the additional mass we need to close the universe as the "missing mass," but it's not really missing; it may not even exist. We do know, however, that there is considerable invisible matter that does exist. It is called dark matter. Dark matter can be identified in a couple of different ways. One is through the measurement of stellar velocities. If we know the velocity of a star in orbit, say, in the Milky Way galaxy, we can deduce the mass of all the stars between it and the center. Indeed, if we can deduce the speed of a star near the outer edge of a galaxy, we can determine the mass of the galaxy. Astronomers have done this in the case of many galaxies and they have found that in almost all cases there is a lot of mass in the galaxy that is not visible. It is assumed that most galaxies have an invisible halo around them, and there is considerable mass in this halo.

Another indication of dark matter was found early on in clusters of galaxies. We know that most galaxies reside in clusters. Astronomers are able to determine the masses of the individual galaxies and the mass of the overall cluster. They found that when the individual masses were added up, the sum was far short of the mass of the cluster. There was a serious discrepancy which meant that there was a lot of mass in the cluster that we couldn't see.

What form does this dark matter take? One possibility is hydrogen gas. Of the three forms of the gas the best candidate was ionized hydrogen gas. Neutron stars, black holes, dust, rocks, and several other things have also been considered. But the bottom line is that most are not considered to be good candidates. Black holes in a certain mass range are a possibility. Radiation is another possibility (it is a form of mass), but calculations show that it is not a good candidate. Various types of exotic particles have been considered, but most are not satisfactory. At the present time we're still not certain what form the dark matter takes, but considerable work is still being done on the problem.

Other methods have been used in our attempt to determine the fate of the universe. In theory, number counts should be useful, but the results have been controversial.[19] You merely have to count the number of galaxies out as far as possible, then make a plot of them at various distances. A flat universe will have a uniform distribution at all distances; a closed universe (positively curved) will have an excess nearby and an open one (negatively curved) will have an excess in its outer reaches.

We are still not sure of the fate of the universe. Most studies indicate that it is open, but barely. And as we saw earlier, inflation theory predicts that it is flat, and will expand forever. Not everyone, however, accepts this.

EINSTEIN AND COSMOLOGY

Our overview of cosmology has been brief and many important points have been left out, but it should give you some idea of the tremendous legacy that Einstein has left us. Not only was his cosmology of 1916 the first modern cosmology, but all (or nearly all) cosmologies that have been formulated since are based on Einstein's general theory of relativity. He is truly the father of cosmology. Most of the important advances have taken place since his death but Einstein eventually accepted the expansion of the universe. He used the cosmological constant in his original theory but in 1932 he discarded it, saying it was the worst blunder of his life. If he had not used it, he may have been able to predict the expansion of the universe. Strangely, even though he discarded it, the cosmological constant continues to be used in cosmology.

Searching
for the Elusive

SOON AFTER EINSTEIN COMPLETED HIS GENERAL
theory of relativity he began thinking about
extending it. General relativity was, after all, an
extension of special relativity. It was natural to
think of generalizing it to explain atoms and elementary particles, but there was also another field
of nature to contend with, namely, the electromagnetic field. Maxwell's theory explained it, but the
more Einstein thought about the field, the more he
was convinced that it was related to the gravitational field. There were, indeed, many similarities
between the two fields. Both depended on sources:
matter was the source of the gravitational field and
charge the source of the electric field. Both had an
infinite range, and both fell off in the same way
with distance. In other words, they got stronger

and stronger as you got closer to the source, and, in theory, they were infinite at the center of the source. We refer to this region as a "singularity."

Furthermore, both fields generated waves in the same way. An oscillating charge produces an electromagnetic wave, and an oscillating mass produces a gravitational wave. At that time, however, very little was known of gravitational waves. Finally, perhaps the most striking similarity between the fields was that they both traveled at the same speed, the speed of light.

Despite these similarities Einstein realized there were significant differences. First of all there was a tremendous difference in their strengths. The electromagnetic field was 10^{37} times as strong as the gravitational field. Furthermore, the gravitational field was generated from one type of source—mass; electric fields were generated from two types—positive and negative charge. As a result, there was both a repulsive and attractive force in the case of electric fields, but only an attractive force in the case of the gravitational field. Another major difference was that you could shield an electric field, but you could not shield the gravitational field.

The differences were, indeed, significant, but the electric and magnetic fields had been unified years earlier, even though at first glance they appeared to be quite different. Light and radio waves (and other electromagnetic waves) had also, in a sense, been unified. No one had suspected, at first, that light was so closely related to radio waves.

Einstein was convinced that there had to be proper generalization of his theory, but strangely, he wasn't the first to publish one.

WORKING IN THE DARK

As Einstein pondered the problem of unification of the electromagnetic and gravitational fields he also wondered about matter, namely, atoms and particles such as electrons and protons. Where did they fit in? As far as he was concerned, matter had no place in a field theory; it would have to come out of the equations of the new theory. He also abhorred the idea of a singularity occurring at the center of matter. And there were the fundamental constants of nature, things such as the speed of light and the charge on the electron. Would they appear naturally in the theory? Charged particles were another problem. Only two particles were known at the time: the electron and the proton. The electron was negatively charged, the proton positively charged. But like charges repel one another, and it was assumed that the electron and proton had a small but finite radius. This meant that one side of the electron would repel the other side. The electron should explode—but it didn't. What was holding the charge together?

Einstein hoped that when he completed his unified field theory it would explain all these problems. In the back of his mind there was, no doubt, the hope that it would be a "theory of everything." Interestingly, though, even before Einstein completed his general theory of relativity, a theory was put forward by Gustav Mie of Greifswald, Germany, that attempted to explain particles.[1] Mie used special relativity along with Maxwell's theory in formulating his theory, but it was soon shown to be lacking. Years later David Hilbert of Göttingen incorporated it into general relativity, but it remained unsuccessful.[2]

In 1919 an ingenious attempt at unification was made by Hermann Weyl of Germany. Weyl had known Einstein for years, and had sent him a draft of his book *Space, Time, and Matter* before it was published. "It is a masterful symphony," Einstein wrote back. Weyl realized that "direction" acted strangely in curved space. Consider, for example, two airplanes at some distance apart near the equator on Earth (see figure 10.1). They are, in effect, on a two-dimensional curved surface. Suppose they both take off and head north. Since they are both heading directly north their paths will be parallel, but during the flight they will get closer and closer together, and when they get to the north pole they will cross over it at an angle to one another. All you have to do is look at the lines of longitude on a globe of the earth to see this. What happened? They took off parallel, and both headed in the same direction throughout the trip, but they ended up traveling at an angle to one another. Direction is obviously not preserved in curved space.

Weyl took things a step further. He assumed that length was

also not preserved. In other words, the two airplanes would arrive at the pole with different lengths if they took different paths. Their shape, however, would be preserved. Adding this to Einstein's general theory of relativity, Weyl obtained extra equations, and when he examined them he found that they were Maxwell's equations of electromagnetism. It was magic! A simple, seemingly natural change and both Einstein's and Maxwell's equations came out of the theory.

Weyl sent his theory to Einstein, asking him to present it to the Prussian Academy and to publish it in the Academy *Proceedings*.

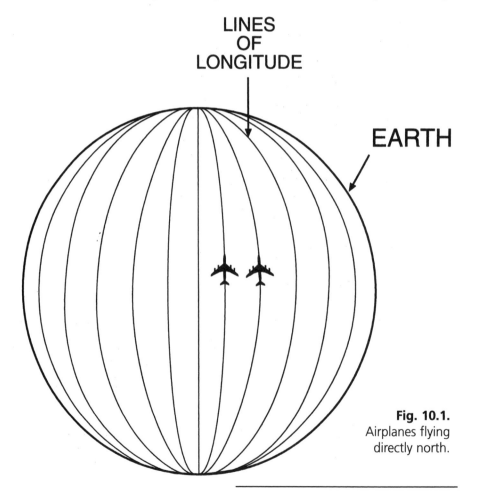

LINES OF LONGITUDE

EARTH

Fig. 10.1.
Airplanes flying
directly north.

Einstein was amazed by Weyl's result. When he studied it in detail, however, he found a flaw. It was no doubt a disappointment to him, and certainly was a disappointment to Weyl. Einstein realized that in taking the "trip" in the curved space you were actually traveling in space-time, not just space, and this meant that time would also change. Moreover, time could be related to vibrational motion, namely, the vibrational motion of an atom. This meant that if two identical atoms took different paths to a given point such as the north pole, they would end up vibrating at different rates. There was no experimental evidence that this was the case; in fact, Einstein was sure it wasn't. He was so impressed with the theory, however, that he sent it to be published, putting his objection in at the end as a footnote. Interestingly, although Weyl's idea failed here, it was later used with considerable success in quantum field theory.[3]

Two years later in 1921 another unified field theory came into Einstein's hands. A mathematician, Theodor Kaluza, of Königsberg, Germany, decided to write down Einstein's equations in five dimensions instead of four. Again, as in Weyl's case, there were extra equations, and when Kaluza analyzed them he found they were Maxwell's equations. How was it possible? Again both Einstein's and Maxwell's equations appeared in the generalization. Kaluza sent his results to Einstein.

Needless to say, Einstein was amazed. Lightning had struck twice and maybe this time it was the "right theory." Indeed, as he checked through the theory looking for a flaw he could find none. But it was strange in one respect. Kaluza had to relate his five dimensions to the four dimensions of the real world. He did this using a "projection"—like the shadow of a three-dimensional object on the wall. To many this was unconvincing. Einstein, however, accepted it and had the theory published.[4]

As others began to study the theory, however, it soon became obvious that there was little in it other than Maxwell's equations. Einstein had hoped that particles would be predicted, but the theory said nothing about them. The theory was made more viable in 1926 by the Swedish physicist Oscar Klein.[5] He suggested that the fifth dimension was not seen because it was "curled up" so tightly it was invisible. Einstein worked on the theory for a while, but eventually lost interest, as did others. Even though the theory was eventually shown to be unsuccessful, many scientists today look back on it as a major breakthrough. It was the first time extra dimensions (other than the usual of four-dimensional space-time) were used, and an ingenious way of explaining why the extra dimension was not visible, was given. As we will see, both of these extensions are used in modern theories today.

Eddington also got into the act. He became intrigued with Weyl's theory and presented a modification of it that he believed got around Einstein's criticism, but it was soon shown to be flawed. Eddington, however, didn't give up, and for years continued searching for a theory of everything. Some felt that in his later years he went too far with what he called his "Fundamental Theory."[6]

EINSTEIN THROWS IN HIS HAT

Although Einstein thought seriously about the problem of a unified field theory for many years, he published nothing. Indeed, he even talked about the possibility in his Nobel address in 1922. His main role at first was as critic. He was impressed with both Weyl's and Kaluza's theories and worked on both for awhile. Both men had generalized his theory. Einstein looked to see if there was another way of generalizing it, one that had been overlooked, and he found one in 1925. General relativity was a symmetric theory in the same way

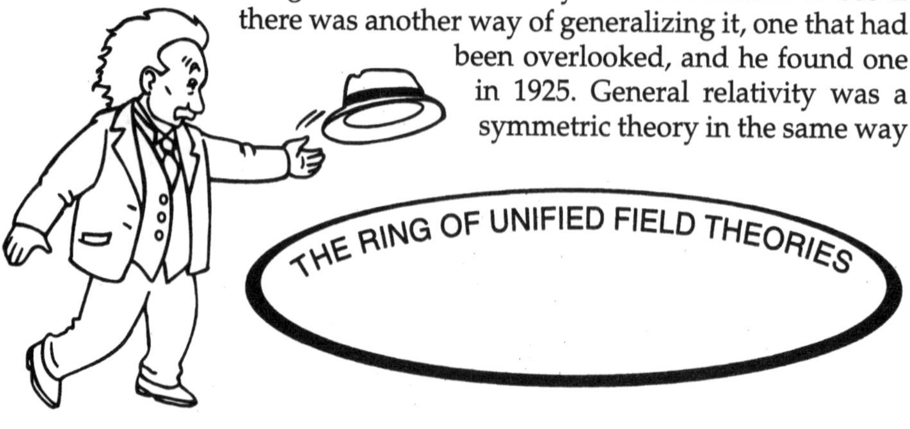

Fig. 10.2. Einstein throws his hat into the ring of Unified Field Theories.

your body is symmetric. Draw a line down the middle of it and look at the two sides; they are the same. He could generalize, or extend, his theory by assuming it was nonsymmetric. The result was a symmetric part and an antisymmetric part, with the symmetric part accounting for general relativity. Einstein examined the antisymmetric part and found it gave Maxwell's equations. He was sure he had at last found the "true solution."[7] This was a natural extension of the theory, but he soon began having difficulties with it. Particles didn't appear out of the equations, as he had hoped. Furthermore, he found what he thought was a serious problem. Looking at the symmetry properties of the theory he found that it predicted mirror-image particles. At this time only the electron and proton were known. This theory seemed to predict that there was a positively charged particle similar to the electron. The proton was positive, but it was 2,000 times heavier than the electron, so it wasn't a mirror-imaged particle. Einstein saw this as a flaw in the theory.

In many ways this is reminiscent of Einstein's problem with cosmology. His theory seemed to predict that the universe was expanding. He added a term to it—the cosmological constant—to stop the expansion. If he hadn't added the term he could have predicted the expansion of the universe. He later called this the greatest "blunder" of his life. In the case of the nonsymmetric theory he found mirror-imaged particles that didn't exist, but a few years later in 1932, Dirac used his relativistic extension of quantum mechanics to predict that such particles did exist. Within a short time after the prediction, the antielectron, or positron, was found, and we know today that all particles have antiparticle partners. Could Einstein have made the prediction from his nonsymmetric theory? It is possible, but he never mentioned it, even after antimatter was discovered.

Einstein eventually left his nonsymmetric theory, although he would later return to it many times. In 1929 he turned back to a variation of Weyl's theory. As he got ready to publish, rumors began to circulate that he had finally "tapped the mind of God." His new theory would explain everything. There were even rumors that it would show scientists how to get to the moon. After all, it would explain the relationship between the electric and gravitational fields, and since electric fields could be shielded, it would show us how we could shield gravitational fields and overcome gravity.

Soon hundreds of reporters were outside his house. He was embarrassed, knowing that the theory hadn't been proved and was, at best, tentative. At first he ignored them, but he had presented the theory to the Prussian Academy and one of the reporters got hold of it and cabled it to New York. The paper appeared in the *New York Times* for everyone to see—equations and all. It's unlikely anyone had the foggiest idea what it meant, but it looked impressive.

The reporters were so persistent that Einstein eventually agreed to an interview, and he selected the reporter from the *New York Times* to conduct it. The reporter, Wythe Williams, brought a photographer with him. Einstein attempted to answer his questions, but it was difficult to explain things in layman terms. Einstein could hardly believe that so much attention was being paid to his theory. When the interview ended the photographer came forward to take his picture. "Do you want me to stand on my head?" Einstein asked jokingly.[8]

The attention was even more of an embarrassment to Einstein several months later when he saw that the theory was flawed and had to discard it. Within a short time, however, he was working on a new theory. Indeed, throughout the rest of his life, as soon as he

---

Einstein in 1938 Referring to his Unified Field Theory

"I still struggle with the same problem as ten years ago. I succeed in small matters but the real goal remains unattainable, even though it sometimes seems palpably close. It is hard and yet rewarding, hard because the goal is beyond my powers, but rewarding because it makes one immune to the distractions of everyday life."[9]

saw that a theory was flawed he would cast it aside and begin working on another.

In 1932 Einstein came to the United States, settling at the Princeton Institute for Advanced Study. He continued working on his theory here; over the years he had many different collaborators. Indeed, he continued his search right up to the day of his death. During this time he appeared to most people to be in an ivory tower, cut off from physics and the physics community. While everyone still respected him and his opinions, there was bewilderment and skepticism about his quest, and there was considerable derision. To many he seemed to have lost touch with the real goals of theoretical physics.

WHAT WAS THE PROBLEM?

Why did Einstein not succeed? After all, he had been so successful in his early life with his other theories. When Einstein started his quest, it was a reasonable one. Only two fields of nature were known—the gravitational field and the electromagnetic field—and two particles, the electron and the proton. He merely wanted a theory that would "unify" and explain them. But as the years passed physics became much more complex. First, quantum mechanics was discovered in the late 1920s. It was extremely successful in explaining the properties of particles, but it was based on uncertainty, probability, and chance, and Einstein abhorred this. To him it was like playing dice with the universe. He was sure it was not the final theory.

But if quantum mechanics was correct, and from all indications it was, it meant that general relativity and quantum mechanics would have to be unified. Einstein was sure that there was a serious problem with quantum mechanics, but it was soon shown that it was general relativity that eventually broke down, not quantum mechanics. We saw in a earlier chapter that general relativity breaks down at extremely high temperatures and densities. In particular it cannot explain the very earliest events of the Big Bang. For years scientists have tried to bring the two theories together. In effect they have tried to find a quantized version of general relativity, but none has ever been found.

In addition to this, as Einstein worked on his unified theories, two other fields of nature were discovered: the weak and strong nuclear fields.[10] Furthermore, large numbers of new particles were discovered; they eventually became so numerous that a new approach was needed to explain them. It gave birth to new, even more elementary particles that we now call quarks.[11] The world was much more complex than Einstein realized. For the most part Einstein ignored these discoveries. Even after it was known that there were four fields of nature, he continued to concentrate on the gravitational and electromagnetic fields. Perhaps he hoped that the other fields would appear in some miraculous way in his equations, but he said little about them.

There was also a problem in relation to Einstein's "game plan." When he was formulating general relativity he had a firm plan in mind, and he knew what he needed. The theory had to explain things that Newton's theory did not, such as the anomaly in Mercury's orbit, and in a first approximation it had to revert to Newton's theory. In searching for his unified field theory he had no such guides. He was on his own.

Finally there was a problem with Einstein's approach. When young he had a certain disdain for mathematics, looking upon it as nothing more than a tool. His appreciation for its powers became evident as he formulated general relativity. But still, he relied mostly on intuition, logic, and thought. Physical insight was important to him at this stage, but strangely, as he aged he relied more and more on mathematics and less on physical intuition. Some viewed this as a fatal flaw.

FINAL REALIZATION

During the 1940s and 1950s Einstein's efforts were looked upon by most with skepticism. He was attempting the impossible, and was not using the right approach. He was working on the wrong problem. He even admitted that to others he must look like an "ostrich with its head buried in the sand." In reality, however, he was years ahead of his time in what he was attempting to do, and scientists gradually began to realize this.

If progress was to be made in physics the various fields of nature and the particles would, indeed, have to be unified. Einstein had shown that gravitation could not be easily unified with electromagnetism. But what about the other fields? Could they be unified? The other fields were quantum fields, however, and a different approach would be needed. The attraction or repulsion between two particles was not a result of a classical action-at-a-distance fields; it was a result of the absorption and emission of particles.

The first theory that incorporated this approach was called quantum electrodynamics, or QED for short. Three scientists came up with it independently about the same time: Richard Feynman of Caltech, Julian Schwinger of Harvard, and Shin'ichiro Tomonaga of Japan. In this theory electromagnetic interactions were a result of the absorption and emission of photons. We know that if two electrons approached one another they repel one another because they have the same charge. In QED this is viewed as the back and forth transfer of photons between them. The more photons exchanged, the greater the force.

QED worked so well a twenty-eight-year-old physicist in Japan, Hedeki Yukawa, became convinced that it could be extended to the strong nuclear force, or strong interactions. The exchange particle would have to be different, but the mechanism would be the same. Yukawa found that the exchange particle would have a mass (unlike the photon, which is massless) about 200 times that of the electron. About ten years later the exchange particle, now called a meson, was discovered. Yukawa went on to show that a similar mechanism would work for the weak nuclear interactions. The exchange particle in this case is called the W particle.

Getting back to our discussion of unification, scientists eventually began looking into the possibility of unifying the electromagnetic field and the weak field. One of those who began working on the problem was Steven Weinberg, who was then at MIT. When Weinberg began there was a serious problem: the photon (the exchange particle of electromagnetic interactions) was massless, but the W particle (exchange particle of weak interactions) had mass. Yet these two particles had to be closely related, in essence, they had to be the same particle. Luckily, within a short time an important breakthrough came in England. Peter Higgs of the University of Edinburgh and a colleague showed that exchange particles could acquire mass by "eating" a strange, new particle, now called a Higgs particle. From a simple point of view the photon could eat a Higgs particle and become a W particle. Weinberg incorporated this in his theory, and Abdus Salam of Trieste came up with the same theory about the same time. But a serious problem remained. Fortunately, it was overcome in 1971 by Gerard t'Hooft of the University of Utrecht, and scientists finally had a successful unification of the weak and electromagnetic fields."

The next step was to join the electroweak theory to the strong interactions. The theory of strong interactions was now much better understood, but it had also become much more complex. Quarks were now "colored," but this isn't color as we usually think of it. They are referred to as red, green, and blue, but it's only a figure of speech. Furthermore, the two major families of elementary particles were now quarks (which made up most of the heavy particles of nature, e.g., protons and neutrons) and leptons (light particles such as the electron). The new theory, which was mainly the brainchild of Murray Gell-Mann of Caltech, was called quantum chromodynamics, or QCD for short. Within the theory there were six different quarks, an exchange particle called the gluon, and six different leptons. The quarks had odd names such as up, down, strange, top, and bottom. The six leptons were the electron and two heavier cousins called the muon and the tau along with their associated neutrinos.[13]

HE'S GOT GUTS

To unify electroweak theory with QCD, in other words, electroweak forces with strong nuclear forces, scientists would have to unify, or bring together, the quark and leptons families. In short, they would have to show that they are part of one and the same

Fig. 10.3. "Hm . . . GUTs . . . it sounds like an interesting theory. What's it about?"

family. This meant that quarks would have to change into leptons and leptons into quarks. Showing this, however, would not be easy. The first attempt was made by Howard Georgi and Sheldon Glashow of Harvard in 1973. We now refer to it as grand unified theory, or GUTs for short. It was a five-dimensional theory with five basic particles. They were three quarks of different colors, the electron and the positron. There would, of course, have to be several exchange particles. In particular, there was one called the X-particle that caused a lepton to change into a quark and vice versa.[14]

With the introduction of the X-particle it was soon evident that the proton could not be stable. It would decay to lighter particles.

The decay time, however, would be exceedingly long—of the order of 10^{31} years. Experiments were set up to check on the decay, but so far it has not been detected.[15]

The Georgi-Glashow theory is now known to be incorrect, but several other GUTs have been put forward. All of them are more complex and the outlook for them at the present time is poor.

STRINGING ALONG

With problems mounting for GUTs, scientists soon began looking beyond it to unification with the last field, gravitation. With gravity in the fold, there would be a complete unification of all fields of nature. Furthermore, it was possible that some of the problems of GUTs and QCD would be solved in a more complete theory. It was well-known, however, that gravity was going to be extremely difficult to incorporate. It was quite different from the other three fields. The only successful theory of gravity was general relativity, and it was a classical theory where gravity was seen as a curvature of space. The other three fields were quantum fields, each with its own exchange particle. It was easy enough to picture gravity as a quantum field; all you had to do was postulate an exchange particle. It had, in fact, been called the graviton. But for it to have any meaning you need a quantum theory of gravitation, and that didn't exist.

After numerous attempts to bring all the fields together, scientists finally realized that an entirely new approach was needed. The idea that "strings" might be important in physics had been around for years, and in 1974 John Schwarz of the University of California and Joel Scherk of France published a paper suggesting that a theory based on strings might qualify as a theory of everything. If so, it would include gravity. There were, however, problems with the theory as it stood, and very few people took the announcement seriously.

In string theory, strings rather than point particles are assumed to be the most elementary things in the universe. All particles and forces are made up of tiny strings. Theories based on strings created some interest in the 1970s, but few people worked on them. One of the few was John Schwarz; in 1979 he went to CERN, sure that he was one of the only people in the world working on strings. To his surprise he met another string enthusiast, Michael Green, and they decided to work together. Both men knew that they would have to overcome many problems if they were to make string theory viable, but they felt confident there was something to the theory.

Several problems confronted them at this stage. One was "infinities" in the theory—places where the theory blew up. Another was "anomalies." A theory with anomalies would not satisfy conservation laws and crazy things could happen. They would have to get rid of them. They worked for two years with little success, but they persevered. Plodding on, they discovered that certain versions of the theory were free of anomalies and infinities. In August 1984, they made their announcement. They had found a theory that appeared to predict and explain all the particles and forces of nature, and it was free of anomalies. The scientific world suddenly began to take an interest. String theory might be the "Holy Grail" of physics, after all. But the theory was still far from problem-free.

One of the people who became interested was Ed Witten, who is now at the Princeton Institute of Advanced Study. He verified

Fig. 10.4. Einstein "stringing along." What would he have thought of string theory?

Schwarz and Green's result and went on to study the theory in detail. Many others were soon attracted to the field and by 1994 there were five different string theories, all of which seemed to be viable. But all of them had problems. In 1995 Witten startled the scientific world by showing that all five theories were different approximations to a deeper, underlying theory. He called it M theory.

There are, however, difficulties even with M theory. First of all, the strings are incredibly small, of the order of 10^{-33} cms. This is 100 billion billion times smaller than the nucleus, about the size of the atom as compared to the solar system. It is so small it is unlikely that any experiments can ever be devised to check the theory. The theory may be unprovable, which is a serious roadblock. Furthermore, the strings are in eleven dimensions. As in the Kaluza-Klein theory, the connection to our world is through "com-

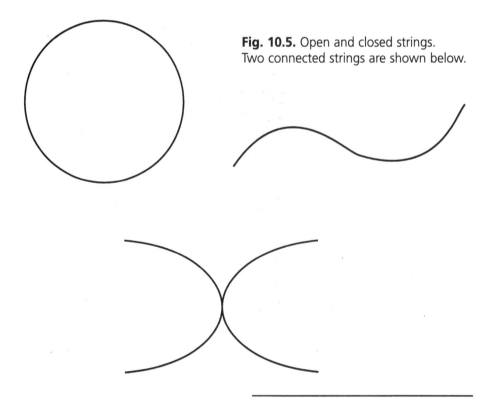

Fig. 10.5. Open and closed strings. Two connected strings are shown below.

pactification." In other words, some of the dimension are curled up so small that we cannot see them.

The strings themselves can be open or closed, and it is assumed that they vibrate at certain frequencies, with waves moving along them in some cases. More recently, vibrating membranes have also been incorporated into the theory. One of the predictions of the theory is a new class of particles called "supersymmetric" particles. We are on the verge of being able to create them with our large accelerators, and they may be discovered in the next few years. If so, they would be very helpful to the theory.

THE FUTURE

Will string theory, or M theory, eventually give us a theory of everything? No one knows at this stage. It definitely has promise, and at the present time it seems to be our best candidate. Witten and others, however, point out that it may take decades before we find out if it is successful. There is much work to be done. One of the major hurdles may be that it appears to be untestable. But science takes unexpected turns, so we can't rule anything out.

11 Quantum Quandary

I T MIGHT SEEM STRANGE THAT Einstein was so strongly against quantum mechanics after having contributed so much to early quantum theory. Many of the older generation, of which Einstein was now part, looked upon the young physicists that formulated the new theory as radicals. Yet there is no doubt that Einstein himself was a radical in the early days of the "older" quantum theory. He admitted that his work on the photoelectric effect, for which he later received the Nobel Prize, was revolutionary.

The "older" quantum theory was born in 1900 when Max Planck of the University of Berlin put forward the idea that radiant energy was emitted in small discrete bundles.[1] He called the bundles "quanta." But discreteness was in conflict with

Maxwell's theory where radiation was seen as continuous, and few accepted the strange idea at first, even though it resolved one of the greatest puzzles of the time: the relationship between the intensity of radiation and its frequency. Einstein, however, didn't hesitate; not only did he accept it, he took it a step further. He postulated that light was an energy particle in the same way as atoms were particles of matter. As justification he pointed to an experiment that had just been performed by Phillip Lenard of Germany. Lenard had shown that electrons were released from an irradiated metal in a strange way. The energy of the released electron was independent of the intensity of the radiation, which didn't appear to make sense. It depended on frequency, and there was a cuttoff frequency below which no electrons were ejected. The phenomenon was called the photoelectric effect.

Einstein tried to make sense of the effect using his idea of light quanta. He began by assuming that the electron absorbed a quantum of light (a light particle) and gained energy. If so, it would be ejected from the metal if the frequency was high enough. But it would likely undergo a number of collisions before it got to the surface, and as a result it would lose energy. Furthermore, there had to be a minimum energy for it to break through the surface—a kind of "work function."[2] It was a strange explanation but it appeared to account for the phenomenon, and it was eventually shown to be correct.

Einstein continued to work on problems related to quantum theory in the years following his paper on the photoelectric effect, but between 1911 and 1916 most of his time was taken up by his general theory of relativity. His thoughts, however, were never far from atomic phenomena. In 1916 he used the quantum concept again to explain the emission and absorption of radiation. He was, in fact, able to derive Planck's radiation formula (the formula that had started the quantum revolution) in a new and interesting way.[3] But Einstein was not satisfied; he had reservations about the method. Radiation was not emitted from an atom in a predictable way; there was an element of chance involved that he didn't like.[4]

Although Einstein was convinced that quanta existed, he was disturbed by the duality that it introduced into physics: radiation was seen as both a particle (later called a photon) and as a wave. How could light be both a wave and a particle? It didn't seem possible. He hoped that both interpretations would eventually be brought together in a more complete theory.

In 1924 a young Indian physicist S. N. Bose of the University of Dacca sent Einstein a paper in which he had been able to derive Planck's formula by treating radiation as a gas made up of particles (photons). His approach was different, however, in the way he counted particles. The standard procedure at the time was to assume they could be individually identified, an idea put forward by Boltzmann many years earlier. Bose argued that photons were indistinguishable (as if they were all red billiard balls) and could not be labeled.

Einstein was delighted with Bose's idea and soon extended it to atoms and molecules. The result was a new form of statistics. Previously everyone had used Boltzmann statistics in dealing with large numbers of particles. Now there were new statistics—Bose-Einstein statistics. Einstein went on to use the new statistics to show that matter would act strangely at very low temperatures.

In 1924 Paul Langevin, a friend of Einstein's from Paris, sent him a thesis that had been submitted to him by one of his students, Louis de Broglie. De Broglie had made the strange proposal that not only photons, but all material particles, behave as waves. Langevin didn't know what to make of the idea and asked Einstein to look it over to see what he thought of it. Einstein was enthusiastic and urged Langevin to accept the thesis. As it turned out the idea was valid; de Broglie won the Nobel Prize for it a few years later. Einstein was, in fact, so enthusiastic about the idea that he put out a challenge to experimentalist in September of 1924 to look for the waves.[5] They were verified in 1927 when Clinton Davisson and Lester Germer in the United States showed that electrons exhibited wave characteristics.

Given this background it does, indeed, seem strange that Einstein was reluctant to accept the new quantum theory that arose over the next few years—a theory we now call quantum mechanics. Until now the theory of the quantum was unconnected and incomplete—a kind of "patchwork" of ideas. Suddenly, however, in the summer of 1925 the first inklings of a unified theory appeared. Einstein heard of it in July and his first reaction was negative. A young physicist in Germany, Werner Heisenberg, had put forward a theory that was so different no one knew what to make of it. But strangely, it worked. At first, even Heisenberg was confused about the mathematics he was using. He introduced "arrays" of numbers and a strange way of multiplying certain quantities, where the order of the terms when they were multiplied mattered (in other words, a × b was not equal to b × a). Furthermore, the theory was based on probability; it didn't give exact predictions, only probabilities. This was an aspect Einstein particularly disliked.

Heisenberg had submitted the theory to his thesis director, Max Born, at Göttingen. Born, who had a strong mathematical background, soon realized that Heisenberg's arrays of numbers were not as new and strange as first thought. They had been used by mathematicians for years, and were known as matrices. The theory began attracting considerable attention when Wolfgang Pauli of Germany showed that it could describe the hydrogen atom better than Bohr's theory. Within months there was a tremendous interest in it.

A conference was organized in December, 1925. It took place in Leyden, with both Einstein and Neils Bohr of Denmark in attendance. Bohr was, by now, a strong supporter of the new theory. Einstein, on the other hand, had reservations and the two men spent many hours discussing and arguing about it. It was the beginning of a friendly feud that would last for years.

Few people were familiar with matrices at the time and to most people Heisenberg's method seemed a little like "black magic." Despite its successes they didn't like it. Then to everyone's surprise another quantum theory was discovered by a theorist in Zurich named Edwin Schrödinger. Prior to formulating his theory, Schrödinger had given little thought to quantum theory, and oddly enough his interest came about by accident. Faculty at the two institutes in Zurich—the Swiss Federal Institute of Technology and Zurich University—had organized a joint colloquium. It was presided over by Peter Debye. News of de Broglie's discovery had made its way to Zurich and everyone was interested in hearing more about it. Debye asked Schrödinger to prepare a colloquium on it. Schrödinger knew nothing about the discovery, but agreed to look into it. He found out as much as he could and presented it at the next colloquium. At the end of the lecture Debye said, "Schrödinger . . . you talk of waves, but you have no wave equation."[6]

It was a simple remark, but it made Schrödinger think. How could he obtain a wave equation? What would it look like? He began looking for one immediately, searching at first for a relativistic equation. He failed to find a satisfactory one, but soon found a nonrelativistic one, and on January 26, 1926, he sent the first of several classic papers to *Annalen der Physik*. Altogether there were five major papers that appeared at roughly one month intervals. These five papers were eventually gathered into a book that was an excellent introduction to the new science of quantum mechanics.

Schrödinger's approach was quite different from Heisenberg's. It was based on differential equations, and differential equations had been used for years in physical theory, so everyone was familiar with them.[7] Einstein was extremely enthusiastic about the theory; it was in a language he was comfortable with. Not only did Schrödinger solve most of the main atomic problems in physics, he also did something else that was vitally needed: he showed that his differential equation method and Heisenberg's matrix method were equivalent. When it came down to fundamentals there was no difference and both theories gave the same results.

There was, however, a serious problem with Schrödinger's approach. The central element was a wave function called psi; it represented the electron (or other particle) in some way, but Schrödinger was unsure how, and there were problems with it.[8] One of the problems was that the wave dispersed and expanded as time passed, yet it was supposed to represent a point particle. How was this possible? Schrödinger struggled to explain it, but wasn't able to come up with anything. Finally Max Born came through with what seemed to be a reasonable explanation: the square of psi gave the probable position of the particle. This is the explanation that we accept today.

UNCERTAINTY AND REALITY

One of the consequences of Heisenberg's theory was that there was a "fuzziness" at the atomic level. If you focused in, for example, on the position of an electron, you could not simultaneously determine its momentum to the same accuracy. Similarly, if you focused in on the momentum, determining it very accurately, you could not simultaneously determine the position to the same accuracy. There is a similar relationship between energy and time. This fuzziness became known as Heisenberg's Uncertainty Principle.[9]

Bohr and Einstein had many friendly arguments over the meaning of quantum mechanics, usually with Einstein presenting a "thought experiment" that showed an apparent flaw, or contradiction, and Bohr showing how it could be overcome. Einstein presented one of these experiments at the Solvay conference of October, 1930. Imagine a box that is filled with radiation, he said. Assume it has a tiny shutter that is opened and closed by a clock. If you weigh the box before and after the emission of the radiation you can obtain the energy content and the time accurately, contrary to the uncertainty principle which states that energy and time cannot be simultaneously measured with high accuracy.

Bohr spent a sleepless night worrying about the conflict but finally came up with an explanation that was an embarrassment to Einstein. He used Einstein's own theory to refute it. According to general relativity the uncertain elevation of the clock during weighing limited the accuracy of the clock, and therefore the uncertainty in time and energy given by the uncertainty principle was valid.

Bohr's position in relation to reality and the measurement process in quantum mechanics eventually became known as the Copenhagen interpretation. According to Bohr, there is no reality unless we measure it. In other words, what exists in the physical world depends on how we measure it. We do not usually think of nature this way. To most of us, the world outside us exists, regardless of whether we measure it or not. According to the Copenhagen interpretation this wasn't true, and as you might expect it caused considerable controversy in the scientific world. Einstein found it difficult to accept and many of his arguments with Bohr centered around it.

Fig. 11.1. Bohr and Einstein.

Bohr also made another important contribution to quantum mechanics called the complementarity principle. It was needed to explain the wave-particle duality of light. According to it the wavelike character and particle-character of light are both mutually exclusive and complimentary. In other words, one excludes the other, but both are necessary to understand light. Light may behave as a particle in one experiment and as a wave in another,

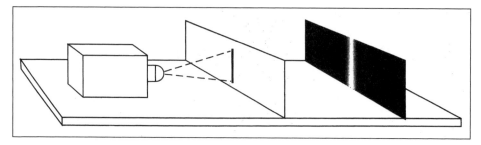

Fig. 11.2. Pattern from a single slit.

but there is no problem if we assume complimentarity. To see why, consider an experiment with a single slit and a double slit. Assume that we begin by directing a beam of photons at the single slit. Most of the photons will pass through as if they were particles; if we set up a screen beyond it we will see a single, symmetric pattern with the greatest intensity at the center—an image of the slit (see figure 11.2). If we now project the beam toward the double slit we will find that we get several regions of high intensity, in other words, several dark and bright regions on our screen (see figure 11.3). It is referred to as an interference pattern. This might seem strange; the results from the single slit would lead us to expect two bright regions, one for each slit, but we get many.

Now, let's reduce the intensity of the beam so that we have single photons going through the slits. For convenience we'll call the two slits A and B. The individual photons have to go through either A or B. If we let them go through the single slit one at a time,

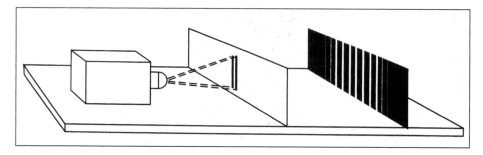

Fig. 11.3. Pattern from a double slit.

eventually we get the same pattern as we did before, namely, a single bright line. Similarly, if we let them go through the double slit, again one at a time, after a large number have gone through we see a number of bright lines, as we did before. This is strange. After all, a given photon goes through only one slit, either A or B. How can we get the same pattern? For this to happen, a photon going through A would have to know if slit B is open or closed. Somehow it has to go over and check to see if it is open; if it is, it goes to one place on the screen and if it is closed it goes to another. But how could a point photon that is traveling directly from the source to the screen check on the second slit? This is where complimentarity comes in. The photon is also a wave, and a wave can spread out. Indeed, we now believe that the wave packet spreads out and senses the second slit, thereby determining where it should go.

We could, of course, set up a small detection device near the slits and check to see which one it actually went through, and if it did, indeed, check the second slit. If we do, however, we disturb it and the uncertainty principle comes into play. If it was going to go through A and we measure it, it might now go through B. So we're back where we started. All in all it's a weird situation, and it was this weirdness that bothered Einstein.

Richard Feynman of Caltech also looked at this experiment and came up with a different explanation.[10] He came to the conclusion that each photon actually goes through both slits. In fact, it does much more than this. He believes that each photon, in traversing the path from the source to the screen, actually traverses every possible path; furthermore, it does so simultaneously. In his view the final result is a combination of all possible paths. It is strange enough that a photon could simultaneously take two different paths, but according to Feynman it was taking an infinite number of different paths, and the method seemed to work.

THE EPR PARADOX

The first paper that Einstein published after coming to the United States in 1933 was a joint one with a twenty-five-year-old Amer-

---

Famous Quotes by Einstein

"God is subtle but He is not malicious."
Interpretation: It might be difficult to find the laws of nature, but it isn't impossible.

"God does not play dice with the universe."
Interpretation: It is not likely that the laws of nature are based on probability.

"Did god have any choice in creating the universe?"
Interpretation: Is our universe the only one possible?

---

ican, Nathan Rosen, and Boris Podolsky, who had just come to the Institute of Advanced Study from Caltech. The paper was titled, "Can the Quantum Mechanical Description of Physical Reality be Regarded as Complete?"[11] The problem the three men presented was called a paradox, but it wasn't really a paradox; Einstein believed he was pointing out a shortcoming of the theory. To him it showed that the physical reality described by quantum mechanics was incomplete.

In the thought experiment Einstein and his colleagues considered two electrons that had just collided. For convenience we will call them electron 1 and electron 2. Assume that after the collision they fly apart and are soon thousands of miles apart. If we now measure the momentum of electron 1, by conservation laws we will know the momentum of electron two. Thus, we have established the momentum of electron 2 without disturbing it in any way. We could have, of course, measured the position of electron 1, rather than its momentum. We would then know the location of electron 2. Note that the measurement of electron 1 will modify the previous measurement of momentum but will not alter the momentum of electron 2—it is far away. This means that the "reality" of the momentum and position of electron 2 depends on a measurement that took place on electron 1, and this is in conflict with the Copenhagen interpretation, which says that reality occurs only when a measurement takes place directly on the system. In this case we know the momentum and position of electron 2 exactly.

The only explanation of this was: either "local causality" was violated or quantum theory was incomplete. Local causality is the idea that events far from one another don't directly influence one another. It doesn't seem possible that it was violated, and according to Einstein and his colleagues this implied that the theory was incomplete.

There is another way of looking at his experiment, one that was formulated by David Bohm of the University of London. Particles have a property called spin. From a simple point of view we can think of them as little spinning tops with their spin axis oriented in a particular direction. Let's assume we have a two particle system with overall spin zero. This means the spin of the two particles have to cancel; this occurs if one has spin up and the other has spin down. Suppose now that the two particles of this system separate with one particle going in one direction and the other going in the opposite direction. We wait until one of the particles (call it particle 2) is thousands of miles away then we measure the spin of particle 1, finding it to be spin up. This tells us immediately that the other particle has spin down. We have determined this even though we have made no measurement on particle 2. But it turns out that we can change the spin of particle 1 by letting it pass through a magnetic field, and if we do, particle 2 still has to have the opposite spin. In other words, the measurement process on particle 1 is somehow affecting particle 2, even though it is thousands of miles away. Is this possible? As I said earlier, the only alternative is that the theory is incomplete, and Einstein was sure this had to be the case.

SCHRÖDINGER'S CAT

Einstein had few allies in his crusade against the philosophical implications of quantum mechanics. Surprisingly, though, Erwin Schrödinger, who had formulated wave mechanics, was on his side. He, too, was disturbed by its weirdness, and he applauded

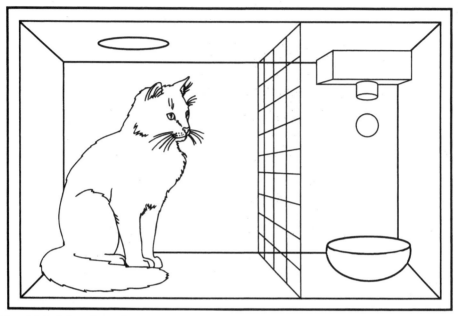

Fig. 11.4. Schrödinger's cat.

Einstein and his colleagues for pointing out the paradox, then went on to present one of his own. It is now referred to as the Schrödinger cat paradox.[12]

Before discussing it let's consider how a quantum "system" is represented. In quantum mechanics a system (which usually consists of some particles) is represented by a wave function. We referred to it earlier as psi. It describes the particles of the system and how they change in time in terms of probabilities. The act of measurement "collapses" this wave function, giving definite values of position, momentum, and so on. It is important to note, however, that it is the act of measurement that gives it these properties. According to quantum mechanics they don't exist before the measurement.

Now consider Schrödinger's cat. In this thought experiment a cat is sealed in a box with a radioactive source and a detector that can detect radioactive particles. We assume that the radioactive source will emit a detectable particle with a probability of 50 per-

Fig. 11.5. "It's Schrödinger's cat."

"It's Schrödinger's cat."   "Is it alive or dead?"   "I don't know."

cent over a period of one minute, and if a particle is detected, a poisonous gas will be released that will kill the cat. Assume further that the box containing the cat is a long distance away from us.

Now turn on the radioactive source for one minute. Is the cat dead or alive at the end of the time? According to the Copenhagen interpretation we cannot tell. As in the case of interacting particles, we have a system that is described by a wave function, and until we collapse it by measuring the system, the cat does not acquire a definite state. Before observation it is neither dead or alive. That's crazy, you say. It has to be one or the other. Not according to quantum mechanics. Until it is actually observed it is neither; it is only a wave function with two equal probabilities. I'm sure you'll agree that this is "weird."

BELL'S INEQUALITY

Scientists continued to argue about the above paradoxes for years. The argument seemed to be that there was an underlying theory beyond quantum mechanics. In other words, quantum mechanics had "hidden variables." This would mean there was a subquantum theory that gave additional information about the state of the system. Bell, a theorist at CERN, attacked the problem directly in 1965. Most of the arguments up to then had been in the form of

Fig. 11.6.
"Which institute
do you
belong to?"

thought experiments. Bell suggested that a "real" experiment could settle the argument. Bell's mathematics took the form of an equality, an equation that could be checked directly by experiment.[13]

The first experiments were inconclusive, but an experiment was performed in 1983 at the University of Paris that was hard to refute. Alain Aspect showed clearly that the inequality was violated and this meant that Bohr was right.[14] Einstein's assumption that the theory was incomplete was not true. There were no hidden variables. This was quite a shock to most physicists. Einstein had assumed that the EPR paradox showed that the theory was incomplete. After all, the only alternative was that local causality was violated. Bell showed that the theory was complete, and this meant that local causality *was* violated. This, in turn, meant that there was some sort of strange "connectedness" between the two particles in the system, and it was instantaneous. But this defies relativity. For the two systems to be connected they must be able to communicate with one another, and the fastest way they could do this is via electromagnetic waves, which travel at the speed of light. If the second particle is millions of miles away the communication can't be instantaneous. Yet the results of Bell's theorem indicate that it is. Again, things are weird.

CONCLUSIONS

Quantum mechanics makes extremely accurate predictions that agree with experiment. The numbers that come out of the theory aren't a problem. But, as we have seen, the philosophical implications of the theory are mind-boggling. First of all, there appears to be a strange connection in the universe that we do not understand. Furthermore, there is a serious problem in relation to reality. Quantum mechanics implies that a system doesn't exist unless we measure it. Most people wouldn't object too strenuously about this in the microworld. In other words, they wouldn't object to the statement that an atom doesn't exist unless we measure or observe it. It's weird, but acceptable perhaps. In theory, though, this idea extends to the macroworld, in other words, to the world we see around us. The dead (or live) cat therefore doesn't exist unless we observe it. That's hard to accept. It means that the mind is a critical part of the universe.

This is, without a doubt, a strange consequence, and we can easily come up with all kinds of dilemmas if it is true. How intelligent, for example, does the mind have to be? If it is observed by a mouse, does it exist? Indeed, if there were no life in the universe to observe it, would it exist? John Wheeler and others have hypothesized that many universes exist, and that life is likely to be rare in them. If there is no life in a particular universe, can we say it exists?

Einstein did not live to see the formulation of Bell's theorem and the outcome. Nevertheless, it's obvious why he was disturbed by quantum mechanics.

12   Epilogue

L OOKING BACK OVER EINSTEIN'S LIFE, IT IS EASY
to see why he is now considered to be one be of
the greatest scientists that ever lived. His legacy is
incredible. His special theory of relativity completely changed our ideas of space and time. He
threw out the idea of an all-pervading ether, and
showed us that all motion is relative and that time
is not the same for all observers throughout the
universe, as had been previously thought.

His theory uncovered many amazing things.
Objects moving relative to an observer are short-
ened in the direction of travel, and they are more
massive. And strangely, at the speed of light, objects disappear and clocks stop. We know, of
course, this can't happen; indeed, it tells us that the
speed of light is unattainable. His theory also

showed that there is a relation between mass and energy. A tremendous amount of energy can be obtained from a small amount of mass. The formula that he derived told us how stars generate their energy, but to Einstein's dismay, it also told us how to build atomic and hydrogen bombs. As a fervent pacifist, he wasn't happy about this.

Einstein's greatest contribution, however, came in 1915. For years he had been trying to generalize his special theory to include acceleration. To his surprise he found his new theory predicted that acceleration and gravity were related. Mass, as the perpetrator of gravity, caused a warping of space-time. Our sun, for example, curved the space around it and the planets traveled in geodesics around this curved space.

Einstein's new theory, general relativity, had even more surprises. Time depended on gravitational fields: the stronger the field, the slower an observer's clock would run, relative to a clock out in space. Soon after it was completed, Einstein applied his theory to the universe, and to his surprise it showed that the universe was unstable. Perplexed, Einstein added a term called the cosmological constant to keep it fixed. He later called this the greatest blunder of his life. Indeed, he could have predicted the expansion of the universe if he had accepted the equations as they were, for it was soon discovered that it was, indeed, expanding.

We now have an exceedingly detailed picture of the evolution of the universe, starting from a tiny fraction of a second after creation through to its final fate. Amazing details have been uncovered, and all of them come from Einstein's theory. But strangely, the theory cannot tell us what happened at the very beginning of the universe, when the density and temperatures were unimaginably high. A quantized version of the theory is needed for this, and no one has found one.

General relativity gave us another surprise. Because clocks run slower on the surface of the earth compared to several miles up in space, a bizarre object called a black hole is possible. In the case of the earth the time difference is negligible, but if the gravitational field is great enough, it can be significant. White dwarfs and neutron stars have very strong gravitational fields and clocks close to their surface therefore run much slower than out in space. Furthermore, there's a surface where time actually stops—the event horizon of a black hole.

Einstein knew his theory was not yet complete. First of all, there was another field of nature, the electromagnetic field, that was not included in it, and it did not explain atoms and elementary particles. It was natural for him to try to extend it, and indeed he did. But it turned out to be a much more difficult challenge than general relativity had been, and he had very little to guide him. To his dismay things didn't come together as they always had previously; it was a tremendous blow to him.

As he struggled to find a unified field theory, as his new theory was called, physics grew more complex. First, a theory of atoms and elementary particles was discovered—quantum mechanics. Then other fields of nature were discovered, and the array of known particles increased considerably. Einstein disliked the new quantum theory almost from the beginning. It gave accurate predictions but it was strange. There was a weirdness about it. Quantum "reality" was quite different from what he was accustomed to: Something was real only if it was measured. Einstein struggled for years to show that the theory was incomplete; he was sure there was a better, more encompassing theory that would overcome the distasteful aspects of quantum mechanics. But he never achieved his goal.

The Best Occupation for a Theorist

Einstein was asked a few years before he died if he would do it all over again, given the chance. He said, "No, I would become a plumber."

On another occasion he remarked that the best job for a theoretical physicist would be as a lighthouse keeper. The peace and quite would be ideal.

On yet another occasion he urged up-and-coming theorists to earn a living as a cobbler to avoid the pressures of publish or perish at a university.

Einstein was still working on his unified field theory when he died on April 18, 1955. He had known for several years that his days were numbered. He had developed an aneurysm in the late 1940s and by 1950 it was beginning to grow. He was told to be careful. To further complicate things, he developed hemolytic anemia in 1954. Still, he managed to hang on. On March 14, he turned seventy-six; several of his friends wanted to give him a birthday party, but Einstein insisted that he didn't want any celebrations.

From a Letter to a Friend

"You and me, though mortal, do not grow old no matter how long we live . . . [because] we never cease to stand like curious children before the great Mystery into which we are born."[1]

Despite problems with his health, Einstein kept busy. In early April he was preparing a speech for the seventh anniversary of the establishment of the state of Israel, and of course, he was working on his unified field theory. Then on Wednesday, April 13, a pain in his chest forced him to bed. His secretary, Helen Dukas, looked after him, but others soon arrived to help. Doctors suggested that he be moved to the Princeton hospital, but he objected. After considerable argument he finally relented and was taken to the hospital on Friday, April 15. The aneurysm had no doubt been perforated, but doctors were sure it could be repaired by surgery. Einstein, however, was adamant. He didn't want his life artificially prolonged. "I would like to go when I want to go," he said. He didn't want to become a vegetable; he preferred to die gracefully.[2]

His friends were frustrated with his stubbornness. Then his son Hans Albert arrived from California, and after Hans talked to him there appeared to be hope that he had persuaded him to undergo the operation. Einstein was now in considerable pain and was being fed intravenously. Surprisingly, though, on Sunday, April 17, he felt better and asked for his notes. He confided to his longtime colleague Otto Nathan that he was very close to success with his unified field theory. That evening he went to sleep peacefully, but shortly after 1:00 A.M. he woke. Nurses rushed in to find him having trouble breathing. They propped him up and put a pillow behind his back. He took several deep breaths, muttered something in German, and died. The aneurysm had burst. When an autopsy was performed it was found that he was in no shape for surgery and would likely have died anyway.

Einstein didn't want a fuss made over him when he died. He wanted no gravestone, no funeral. Nothing. And indeed, he was cremated and his ashes were scattered in the Delaware River. One of his wishes, however, was not carried out. The pathologist, Dr. Thomas Harvey, that performed the autopsy took his brain, despite some objection from Dukas and Nathan. (He did, however, have Einstein's son's permission.) Dr. Harvey kept the brain in a solution of formaldehyde in his house for many years, supplying several people and universities with small sections of it for study.[3]

Was Einstein's brain different in any way from others? Considering his genius you might guess that it would be. But there has been considerable controversy. Dr. Marian Diamond of the University of California at Berkeley says it is different. She looked at the glial cells and said they were more numerous than the average person's in the section of the brain having to do with math. Dr. Rorke of Philadelphia, however, disagreed. She studied several slides of sections of Einstein's brain and believes that Diamond arrived at her conclusions because the sections she studied were a little thicker than usual. Rorke admits, however, that the brain was well-preserved for somebody his age, and there was no sign whatsoever of Alzheimer's disease.[4] In another study, scientists showed that the Sylvian fissure on the surface of Einstein's brain was different from most others. Because of this, his parietal lobes were larger than normal. This is the section of the brain responsible for mathematical intuition. Different or not, it was, indeed, a marvelous brain.

Notes

1. EINSTEIN AS A YOUTH

1. Albrecht Fölsing, *Albert Einstein: A Biography* (New York: Viking, 1997), p. 3.
2. Ibid., p. 35.
3. Ibid., p. 68.
4. Banesh Hoffmann, *Albert Einstein: Creator and Rebel* (New York: Viking, 1972), p. 31.
5. Ibid., p. 77.

2. THE MICHELSON-MORLEY EXPERIMENT

1. Martin Gardner, *The Relativity Explosion* (New York: Vintage, 1976), p. 22.
2. Albert Michelson and Edward Morley, *American Journal of Science* 34 (1887): 333.

3. Ronald Clark, *Einstein: The Life and Times* (New York: World Publishing, 1971), p. 82.

3. SPECIAL RELATIVITY

1. John Stachel, ed., *Einstein's Miraculous Year: Five Papers that Changed the Face of Physics* (Princeton: Princeton University Press, 1998) p. 3.
2. Ibid., p. 11.
3. Fölsing, *Albert Einstein*, p. 120.
4. Vega is one of the brightest stars in the summer sky. It is in the constellation Lyra.
5. This statement is made in the second paragraph of Einstein's paper, "On the Electrodynamics of Moving Bodies," which is the classic paper on special relativity. The experiment is also referred to in a paper by Wilhelm Wien that he is known to have read. See Fölsing, *Albert Einstein*, p. 217.
6. Denis Brian, *Einstein: A Life* (New York: Wiley, 1996) p. 61
7. Double, or binary, star systems are very common in the universe; in fact, most stars are systems of this type. The two stars orbit their common center of mass and are held together by their mutual gravitational attraction.
8. Albert Einstein, "On the Electrodynamics of Moving Bodies," *Annalen der Physik* 17 (1905).
9. For the more mathematically inclined the formula is written as $E = mc^2$, with the square of the speed of light being almost 10^{17} (in metric units) it is obvious that only a few grams of matter is equivalent to a lot of energy.
10. Fölsing, *Albert Einstein*, p. 200.
11. Ibid.

4. FOUR-DIMENSIONAL SPACE-TIME AND TIME TRAVEL

1. Fölsing, *Albert Einstein*, p. 207
2. Brian, *Einstein: A Life*, p. 18.
3. Fölsing, *Albert Einstein*, p. 58.

4. Ibid., p. 243.

5. Banesh Hoffmann, *Creator and Rebel: Albert Einstein* (New York: Viking, 1992), p. 89.

6. Ibid.

7. H. G. Wells, *The Time Machine* (1895; reprint, New York: Tor books, 1995)

8. All of these books are still available in most cases as a reissue. They can be obtained from an internet bookstore such as Amazon.com.

9. Fred Hoyle, *October the First Is Too Late* (New York: Harper and Row, 1966).

10. Helen Dukas and Banesh Hoffmann, *Albert Einstein: The Human Side* (Princeton: Princeton University Press, 1979), p. 31.

5. GENERAL RELATIVITY

1. Brian, *Einstein: A Life*, p. 72.

2. Ibid., p. 71.

3. Nigel Calder, *Einstein's Universe* (New York: Greenwich House, 1979), p. 79.

4. Fölsing, *Albert Einstein*, p. 214.

5. Ibid., p. 283.

6. This is discussed in chapter 7 in the section "Clocks in a Gravitational Field."

7. It is the geometry that was devised by Euclid and published in his book *Elements*. It begins with a number of fundamental definitions, basic postulates, and five axioms. All of the geometry is developed from them.

8. Fölsing, *Albert Einstein*, p. 293.

9. Hoffmann, *Creator and Rebel: Albert Einstein*, p. 116.

10. Ibid., p. 117

11. Fölsing, *Albert Einstein*, p. 315.

12. Roger Highfield and Paul Carter, *The Private Lives of Albert Einstein* (London: Faber and Faber, 1993), p. 167.

13. Fölsing, *Albert Einstein*, p. 375.

14. The entire article is reprinted in *The Principle of Relativity* (New York: Dover, 1923) and in John Stachel, *Einstein's Miraculous Year*.

6. GRAVITY AND CURVED SPACE-TIME

1. Fölsing, *Albert Einstein*, p. 369.
2. Ibid., p. 384.
3. Kip Thorne, *Black Holes and Time Warps* (New York: Norton, 1994), p. 392.
4. Marcia Bartusiak, *Thursday's Universe* (New York: Times Books, 1986), p. 204.

7. TESTING THE THEORY

1. Fölsing, *Albert Einstein*, p. 438.
2. Ibid., p. 439.
3. Ibid., p. 442.
4. Ibid., p. 443.
5. Ibid., p. 444.
6. Ibid.
7. Ibid., p. 439.
8. Vallentin, Antonina, *The Drama of Albert Einstein* (New York: Doubleday, 1954), p. 82.
9. Vulcan was an early Roman god of fire. Since the planet was presumed to be very close to the sun the name was appropriate. It has, of course, never been found, even with modern equipment.
10. It was discovered in 1974. A radio source that flashes seventeen times a second, it is believed to be two neutron stars or perhaps a neutron star and a black hole. The orbital period decreases by about one ten-thousandth of a second a year. As a result the two objects move closer to one about one yard in a year.
11. A white dwarf is a star that was once about as massive as our sun, or slightly more massive. When it ran out of fuel (hydrogen) it collapsed in on itself. About 500 have been discovered. They have a radius of about 10,000 miles.
12. Under usual conditions atoms recoil when they emit radiation such as gamma rays. Mössbauer determined that under special conditions the entire crystal from which a gamma ray is emitted may take up the recoil. In this case the gamma ray is emitted with a very narrow spread win wavelength. This is called the Mössbauer effect.
13. Eric Chaisson, *Relatively Speaking* (New York: Norton, 1976), p. 118.

14. Soon after large numbers of radio sources were discovered in the 1950s, astronomers at Cambridge University decided to catalogue them. Their first catalogue, called C1, had fifty entries. In 1955 they undertook a much more extensive survey, but there were so many problems with false entries that they initiated a third survey soon after it was completed. It was completed in 1960 and was called the 3C catalogue. It is still used extensively by astronomers today.

15. Nigal Calder, *Einstein's Universe*, p. 57.

16. Ibid., p. 38.

17. Brian, *Einstein: A Life*, p. 103.

18. Antonina Vallentin, *The Drama of Albert Einstein* (New York: Doubleday, 1954), p. 84.

8. BLACK HOLES AND OTHER EXOTIC OBJECTS

1. D. Howard and J. Stachel, eds., *Einstein and the History of General Relativity* (Boston: Birkhauser, 1989), p. 216.

2. Kip Thorne, *Black Holes and Time Warps*, p. 134.

3. Ibid., p. 123.

4. A star is stable throughout its life because two forces within it are balanced: an outward force due to the gas pressure created by the thermonuclear furnace at the center of the star, and an inward gravitational force. When the star uses up its fuel the thermonuclear furnace goes out and gravity overwhelms the star causing it to implode.

5. Details of the life of Oppenheimer can be found in Peter Goodchild, *Robert Oppenheimer: Shatterer of Worlds* (Boston: Houghton Mifflin, 1981).

6. Neutron stars occur in the collapse of a star with a mass between 1.4 and 3.2 solar masses. A supernova explosion occurs with the collapse. The neutron star that is left after the collapse is only a few miles across and composed mostly of neutrons. It is exceedingly dense and would likely be spinning rapidly and would have a strong magnetic filed. When pulsars were discovered, they were later identified with neutrons stars.

7. Barry Parker, *Einstein's Dream* (New York: Plenum, 1986), p. 105.

8. A quasar is a radio source that looks like a star but lies well outside our galaxy. According to their red shifts they are more distant than galaxies.

9. The "No Hair Theorem" was a result of the work of several sci-

entists, most notably, Yakov Zel'dovich, John Wheeler, and Dennis Sciama. In essence it says that black holes can only have certain properties. They have been shown to have mass, spin, and charge.

10. According to the story Newman was teaching an undergraduate class on relativity and black holes when he mentioned that a solution to the case where the black hole had both charge and spin had never been found. One of the students pointed out that if the transformation between the ordinary, or Schwarzschild black holes, and Kerr black holes was known, it could be applied to the Reissner-Nordström black hole and it would give a solution for the spinning, charged black hole. Newman assigned it as homework. A brief paper was published shortly thereafter announcing the solution; Newman and all his students had their names on the paper.

11. William Kaufmann III, *The Cosmic Frontiers of General Relativity* (Boston: Little, Brown, 1977), p. 177.

12. Further details can be found in Kaufmann, *The Cosmic Frontiers of General Relativity*, p. 146.

13. A few internet sites on black holes are:
http://www.ast.cam.ac.uk/pubinfo/leaflets/blackholes/blackholes.html
http://www.astronomical.org/astbook/blkhole.html
http://physics7.berkeley.edu/BHfaq.html
http://web.syr.edu/~jebornak/blackholes.html
http://antwrp.gsfc.nasa.gov/htmltest/rjn_bht.html

14. Barry Parker, "Where Have All the Black Holes Gone?" *Astronomy* (October 1994) p. 39.

15. Details of his life can be obtained from many different sources. A particularly good one is John Boslough, *Stephen Hawking's Universe* (New York: Avon, 1985)

16. Stephen Hawking, *A Brief History of Time* (New York: Bantam, 1985) p. 103.

17. William Kaufmann III, *The Cosmic Frontiers of General Relativity*, p. 129.

18. Barry Parker, *Cosmic Time Travel* (New York: Plenum, 1991), p. 197.

19. Kip Thorne, *Black Holes and Time Warps*, p. 483.

20. Ibid., p. 508.

21. Fölsing, *Albert Einstein*, p. 536; Brian, *Einstein: A Life*, p. 143.

22. Ibid.

9. TO THE ENDS OF THE UNIVERSE

1. A galaxy is a large system of stars, usually containing millions to hundreds of millions of stars. Gas and dust are also frequently present. We live in the Milky Way galaxy which is about 100,000 light years across. Galaxies are classified as spiral or elliptical according to their shape and appearance. A few are known as irregulars.

2. Georges Lemaître, "The Expanding Universe," *MNRS* 91 (1931): 490.

3. William Sheehan, *The Planet Mars* (Tucson: University of Arizona Press, 1997), p. 98.

4. The Andromeda galaxy is about two million light years away. It is one of our nearest galaxies and is slightly larger than the Mikly Way galaxy. It has a bright nucleus and long spiral arms and can be seen with the naked eye in the constellation of Andromeda.

5. Cepheid variables area type of variable star. In other words, their light intensity changes periodically. The period of the Cepheids is one to fifty days. The star itself is a supergiant.

6. Edwin Hubble, *The Realm of the Nebulae* (New Haven: Yale University Press, 1936).

7. The best source of information on Gamow is his own book: *My World Line* (New York: Viking, 1970)

8. More exactly, deuterium, tritium, helium, and a little lithium would be produced.

9. They are published in *Astrophysical Journal Letters*. The reference is *Astrophysical Journal* 142, 419 (1965).

10. The Kelvin temperature scale is a scale that was set up by Lord Kelvin the lowest temperature in the universe is $0°$ K. It is $459°$ F.

11. Much more on this topic can be found in a book by one of the major investigators George Smoot. The reference is: George Smoot and Keay Davidson, *Wrinkles in Time* (New York: Morris, 1993).

12. Dennis Overbye, *Lonely Hearts of the Cosmos* (New York: Harper-Colins, 1991), p. 179.

13. The notation here is referred to as scientific notation. A number such a 10,000,000 is referred to as 10^7 according to the number of zeros. Similarly a number such as 1/100,000 is referred to as 10^{-5}.

14. Barry Parker, *Creation* (New York: Plenum, 1988), p. 210.

15. Barry Parker, *The Vindication of the Big Bang* (New York: Plenum, 1993), p. 237.

16. More exactly omega is (average density of the universe)/(critical density).

17. Michael Riordan and David Schramm, *The Shadow of Creation* (New York: Freeman) 1990), p. 117.

18. Our Local Group of galaxies, a group of about twenty-five galaxies that includes the Milky Way, is part of a gigantic cluster of clusters called a supercluster. Our supercluster is called the Local Supercluster. The huge Virgo cluster at the center contains about 2,500 galaxies.

19. Joseph Silk, *The Big Bang* (New York: Freeman, 1980) p. 305.

10. SEARCHING FOR THE ELUSIVE

1. Fölsing, *Albert Einstein*, p. 555.
2. Ibid.
3. Ibid., p. 556.
4. Ibid., p. 557.
5. Ibid.
6. Ibid.
7. Ibid., p. 563.
8. Brian, *Einstein: A Life*, p. 174.
9. Helen Dukas and Banesh Hoffmann, *Albert Einstein: The Human Side* (Princeton: Princeton University Press, 1979), p. 69.
10. Heinz Pagels, *The Cosmic Code*, p. 226.
11. Quarks are point particles that make up all the hadrons (heavy particles) of the universe. They have a fractional electric charge and come in six varieties. These varieties (flavors) are: up, down, strange, charm, bottom, and top. The exchange particle that holds them together is called the gluon.
12. Considerably more information on this topic can be found in the author's book, *Search for a Supertheory* (New York: Plenum, 1987).
13. Ibid.
14. Ibid.
15. Ibid.

11. QUANTUM QUANDARY

1. Pagels, *The Cosmic Code*, p. 26.
2. Ibid., p. 29.
3. Fölsing, *Albert Einstein*, p. 567.

4. Ibid., p. 568
5. Ibid., p. 577.
6. Parker, *Search for a Supertheory*, p. 30.
7. Differential equations are familiar to anyone that has taken calculus. They are the basic equations within that branch of mathematics.
8. Parker, *Search for a Supertheory*, p. 31.
9. Pagels, *The Cosmic Code*, p. 87.
10. Brian Greene, *The Elegant Universe* (New York: Norton, 1999), p. 108.
11. John Gribbin, *In Search of Schrödinger's Cat* (New York: Bantam, 1984), p. 181
12. Ibid., p. 203.
13. Ibid., p. 222.
14. Ibid., p. 228.

12. EPILOGUE

1. Helen Dukas and Banesh Hoffmann, *Albert Einstein: Creator and Rebel* (Princeton: Princeton University Press, 1979), p. 72.
2. Brian, *Einstein: A Life*, p. 426.
3. Ibid., p. 437.
4. Ibid. p. 438.

Glossary

ABSOLUTE MOTION: Motion that is the same regardless of the system it is measured in.

ABSOLUTE TIME: A universal time that is the same for all observers in the universe, independent of their motion.

ABERRATION (OF STARLIGHT): The apparent shift in a star's position due to the earth's orbital motion.

ACCELERATION: The rate of change of velocity.

ACCRETION DISK: A flattened disk of matter spinning around a star or black hole.

ACTIVE GALAXY: Radio galaxy. Galaxy that gives off radio waves and other radiations.

ANISOTROPY: Refers to whether the universe is the same in all directions. If it is not the same it is anisotropic.

ANOMALY: An irregularity.

ANTIMATTER: Matter consisting of antiparticles such as antiprotons.

ATOMIC CLOCK: Highly accurate clock. Mechanism based on atomic phenomena.

AXIOM: A self-evident truth.

BIG BANG THEORY: Theory of the creation of the universe. Assumes universe began as an explosion.

BINARY SYSTEM: A system of two stars held together by their mutual gravitational pull.

BINARY PULSAR: Refers to a particular system of two pulsars in orbit around one another.

BLACKBODY CURVE: The curve that results when the intensity of radiation is plotted against wavelength or frequency.

BLUESHIFT: A shift of spectral lines toward the blue end of the spectrum. It indicates approach.

CAPILLARY ACTION:	The rise of water or other fluid in a narrow or very fine tube.
CAUSALITY:	The principle that says cause must come before effect.
CENTER OF MASS:	The "average" position of a group of massive bodies. It lies between them.
CEPHEID VARIABLES:	A star of varying brightness. Periodic with period between one and 50 days.
CLASSICAL THEORY:	Any nonquantum theory. Newton's theory, Maxwell's theory of electromagnetism, and relativity theory are examples.
CLUSTER:	A group of stars held together by their mutual gravitational attraction.
COMPACTIFICATION:	A "curling up" of a higher dimension.
CONSTELLATION:	A group of stars that appear to be close together in the sky.
CONTINUUM:	A continuous region.
COSMIC RAY:	Radiation and particles from deep space. Very energetic.
COSMOLOGICAL CONSTANT:	Constant Einstein added to his equations of general relativity to stabilize the universe.
COSMOLOGY:	The study of the structure of the universe. Usually includes evolution.

COVARIANCE: Implies that the form of the equations remains the same in any transformation.

DENSITY: Mass per unit volume.

DENSITY FLUCTUATION: A change in density from place to place.

DEUTERIUM: A heavy form of hydrogen. Nucleus contains one proton and one neutron.

DIFFERENTIAL EQUATION: An equation in calculus. Can give time evolution of a system.

DIFFUSION RATE: Rate at which molecules diffuse through a medium.

DOPPLER SHIFT: A change in wavelength that occurs when a body emitting waves is either approaching or receding.

ELECTRIC FIELD: Field around a charged particle.

ELECTRODYNAMICS: The study of the interaction of charged particles.

ELECTROMAGNETIC FORCE: The force that arises between two charged particles.

ELECTROMAGNETIC WAVE: A wave given off by oscillating electric charges.

ELECTRON: Basic particle of electric current. Also component of atom. Negatively charged.

ELLIPSE: An egg-shaped curve.

ERGOSPHERE: Region between the event horizon and static limit of a black hole.

ESCAPE VELOCITY: Velocity required to overcome a particular gravitational field.

ETHER: A hypothetical substance believed at one time to permeate the universe. Needed to propagate electromagnetic waves.

EVENT HORIZON: Surface of a black hole. A one-way surface.

EXCHANGE PARTICLE: Particle that is passed back and forth in interactions, e.g., photon is the exchange particle of the electromagnetic interactions.

EXOTIC MATTER: Matter needed to stabilize a wormhole.

FLUID DYNAMICS: The study of the flow of fluids.

FRAME OF REFERENCE: A system that is used as a standard, to which everything can be compared.

FREQUENCY: Number of vibrations per second.

GALAXY: System consisting of billions of stars.

GEODESIC: The shortest distance between two points. Can also be the longest.

GLOBULAR CLUSTER: A cluster of stars smaller than a galaxy. Usually contains a few hundred thousand stars.

GLUON: The exchange particle of the strong interactions.

GRAVITATIONAL RADIUS: Radius at which the escape velocity is equal to the velocity of light.

GREAT ATTRACTOR: A huge accumulation of mass that has produced a measurable effect on the motion of the Local group of galaxies.

HADRON: Class of particles made up of particles that participate in the strong interactions.

HALF-LIFE: Time for half of a radioactive sample to decay.

HYPERBOLA: One of the open conic curves.

IONIZED GAS (HYDROGEN):   Gas that has lost orbital electrons and is charged.

IMAGINARY NUMBER: A number that is not in the system of real numbers. The square root of a negative number is imaginary.

IMPLOSION: Collapse inward. An inward explosion.

INERTIA: Resistance to change in motion.

INFLATION: Refers to a sudden, very dramatic expansion that may have occurred in the early universe.

INTERFERENCE: Occurs when two rays of light merge. If loops and nodes of the wave line up there is constructive interference. If opposite there is destructive interference.

ISOTROPIC: The same in all directions.

KELVIN TEMPERATURE SCALE: A temperature scale set up with zero degrees as the lowest possible temperature in the universe.

KERR BLACK HOLE: A spinning black hole.

KINETIC ENERGY: Energy of motion.

LAGRANGIAN POINT: Point between two bodies where their gravitational pull is equal.

LASER: Instrument that gives off a coherent beam of light.

LEPTON: Light particle such as the electron.

MAGNETIC MONOPOLE: A heavy particle with either a south or a north pole, but not both.

MASS: A measure of the amount of matter in a body.

MATRIX: An array of numbers used in mathematics.

MESON: a medium heavy particle.

MILKY WAY GALAXY: The galaxy or system of stars that we live in.

MOMENTUM: Mass multiplied by velocity.

NAKED SINGULARITY: Singularity with no event horizon around it.

NEGATIVE CURVATURE: A type of curvature. A saddlelike surface has negative curvature.

NEUTRINO: A massless or very light particle associated with weak interactions.

NEUTRON STAR: A star made up mostly of neutrons. Very dense and compact.

PERTURBATION: A small disturbance or change.

PHOTOELECTRIC EFFECT: The emission of electrons from a metal when light is shone on it.

PHOTON SPHERE: Surface around a black hole. Lies 1.5 times farther out than the event horizon.

PHOTON: A particle or light. The exchange particle of the electromagnetic interactions.

POSITRON: A positive electron. Similar to electron but of positive charge.

PRECESSION: A slow change in orientation of the major axis of an elliptical orbit.

PRIMORDIAL
BLACK HOLE: Black hole created in the Big Bang explosion.

PROTON: Basic component of the nucleus of the atom. Positively charged.

PULSAR: A short-period variable star. Composed of neutrons.

QUANTUM MECHANICS: Theory of atoms and molecules, their structure and interactions with radiation.

QUANTUM CHROMODYNAMICS: Quantum field theory that describes interactions between quarks and gluons.

QUANTUM: A discrete amount of energy that is absorbed or emitted in particle interactions.

QUASAR: A starlike object with a large redshift. Strong source of radio waves.

RADIATION: Electromagnetic energy or photons.

RADIO TELESCOPE: Telescope used for detecting radio sources.

RADIO SOURCE: A source of radio waves. Usually a galaxy or star.

RADIOACTIVE: Gives off radiation and energetic particles.

RECESSIONAL VELOCITY: Velocity away from an observer.

REDSHIFT: A shift of spectral lines towards the red end of the spectrum. It indicates recession.

RING SINGULARITY: Singularity in the form of a ring. The Kerr black hole has a ring singularity.

SCHWARZSCHILD RADIUS: Radius at which the escape velocity is equal to the velocity of light.

SINGULARITY: A region where a theory goes awry and gives incorrect answers. Point or ring in the center of a black hole.

SPACE-TIME: A four-dimensional unification of space and time.

SPECTROSCOPE: Instrument for observing spectral lines.

SPECTRUM: Lines seen when light is passed through a spectroscope.

STATIC LIMIT: Region near a black hole where it is impossible to be stationary.

STATISTICS: Numerical facts systematically collected.

STELLAR SYSTEM: A system of stars.

SUPERCLUSTER: A cluster of clusters.

SUPERNOVA: An exploding star.

TAU: The heaviest known lepton.

TENSOR: Component in complex branch of mathematics called tensor analysis. The equations of general relativity are written in terms of tensors.

THERMONUCLEAR: Refers to nuclear reactions that give off heat in the core of a star.

TIDAL FORCE: Force on a body due to a varying gravitational field.

TIME DILATION: A decrease in a time interval caused by motion.

TRANSFORMATION: A mathematical relation between two systems. A change of coordinates.

UNCERTAINTY
PRINCIPLE:

Principle that states that there is an uncertainty when we attempt to measure various variables in physics simultaneously.

UNIFIED FIELD THEORY:

An attempt to include electromagnetism in general relativity, or more generally, to unify all fields.

VIRTUAL PAIR:

A particle-antiparticle pair that appears briefly out of the vacuum, then disappears.

W PARTICLE:

The exchange particle of weak interactions.

WAVE FUNCTION:

Component used in quantum mechanics. Associated with a particle, particularly the particle's position.

WHITE DWARF:

A small dense star; slightly larger than Earth.

WHITE HOLE:

Associated with the exit end of the wormhole of a black hole.

WHITE NEBULAE:

Early term used to describe white, diffuse objects in the sky.

WORLD LINE:

The line of "events" or position of an observer over time.

WORMHOLE:

Warped space in the form of a wormhole leading up to a black hole.

X-PARTICLE:

Extremely heavy exchange particle predicted by GUTs. They allow quarks to change into leptons and vice versa.

Bibliography

1. EINSTEIN AS A YOUTH

Bernstein, Jeremy. *Einstein*. New York: Viking, 1973.

Brian, Denis. *Einstein: A Life*. New York: Wiley, 1996.

Clark, Ronald. *Einstein: The Life and Times*. New York: World, 1971.

Fölsing, Albrecht. *Albert Einstein: A Biography*. New York: Viking, 1997.

Frank, Philipp. *Einstein: His Life and Times*. New York: Knopf, 1972.

Highsmith, Roger, and Paul Carter. *The Private Lives of Albert Einstein*. London: Faber and Faber, 1993.

Hoffmann, Banesh. *Albert Einstein: Creator and Rebel*. New York: Viking, 1972.

Michelmore, Peter. *Einstein: Profile of the Man*. London: Muller, 1963.

Pais, Abraham. *Subtle Is the Lord*. New York: Oxford, 1982.

Vallentin, Antonina. *The Drama of Albert Einstein.* New York: Doubleday, 1954.

2. THE MICHELSON-MORLEY EXPERIMENT

Chaisson, Eric. *Relatively Speaking.* New York: Norton, 1976.
Gardner, Martin, *The Relativity Explosion.* New York: Vintage, 1976.
Kaufmann, William, III. *The Cosmic Frontiers of General Relativity.* Boston: Little, Brown, 1977.

3. SPECIAL RELATIVITY

Barnett, Lincoln. *The Universe and Dr. Einstein.* New York: New American Library, 1953.
Einstein, Albert. *Relativity.* New York: Crown, 1961.
Chaisson, Eric. *Relatively Speaking.* New York: Norton, 1988.
Gardner, Martin. *The Relativity Explosion.* New York: Vintage, 1976.
Kaufmann, William, III. *The Cosmic Frontiers of General Relativity.* Boston: Little, Brown, 1971.
———. *Black Holes and Warped Spacetime.* New York: Freeman, 1979.
Parker, Barry. *Cosmic Time Travel.* New York: Plenum, 1991.
Thorne, Kip. *Black Holes and Time Warps.* New York: Norton, 1994.
Zukav, Gary. *The Dancing Wu Li Masters.* New York: Morrow, 1979.

4. FOUR-DIMENSIONAL SPACE-TIME AND TIME TRAVEL

Calder, Nigel. *Einstein's Universe.* New York: Greenwich, 1979.
Kaufmann, William, III. *The Cosmic Frontiers of General Relativity.* Boston: Little, Brown, 1977.
Macvey, John. *Time Travel.* Chelsea: Scarborough House, 1990.
Narlikar, Jayant. *The Lighter Side of Gravity.* New York: Freeman, 1982.
Parker, Barry. *Cosmic Time Travel.* New York: Plenum, 1991.
Rucker, Rudy. *The Fourth Dimension.* Boston: Houghton Mifflin, 1984.

5. GENERAL RELATIVITY

Barnett, Lincoln. *The Universe and Dr. Einstein*. New York: New American Library, 1953.
Einstein, Albert, and Leopold Infeld. *The Evolution of Physics*. New York: Simon and Schuster, 1951.
Einstein, Albert. *Relativity*. New York: Crown, 1961.
Gamow, George. *One Two Three . . . Infinity*. New York: New American Library, 1954.
Gardner, Martin. *The Relativity Explosion*. New York: Vintage, 1976.
Kaufmann, William, III. *Relativity and Cosmology*. New York: Harper and Rowe, 1977.
Stachel, John, ed. *Einstein's Miraculous Year*. Princeton: Princeton University Press, 1998.
Thorne, Kip. *Black Holes and Time Warps*. New York: Norton, 1994.

6. GRAVITY AND CURVED SPACE-TIME

Davies, Paul. *The Edge of Infinity*. New York: Simon and Schuster, 1981.
Ferris, Timothy. *The Whole Shebang*. New York: Simon and Schuster, 1997.
Gribbin, John. *Time Warps*. New York: Delacortes, 1979.
Hawking, Stephen. *A Brief History of Time*. New York: Bantam, 1988.
Kaufmann, William, III. *Black Holes and Warped Spacetime*. New York: Freeman, 1979.
Narlikar, Jayant. *The Lighter Side of Gravity*. New York: Freeman, 1982.
Wald, Robert. *Space, Time and Gravity*. Chicago: University of Chicago Press, 1977.

7. TESTING THE THEORY

Brian, Denis. *Einstein: A Life*. New York: Wiley, 1996.
Chaisson, Eric. *Relatively Speaking*. New York: Norton, 1988.
Fölsing, Albrecht. *Albert Einstein: A Biography*. New York: Viking, 1977.
Gardner, Martin. *The Relativity Explosion*. New York; Vintage, 1976.

8. BLACK HOLES AND OTHER EXOTIC OBJECTS

Asimov, Isaac. *The Collapsing Universe*. New York: Pocket Books, 1977.
Boslough, John. *Stephen Hawking's Universe*. New York: Avon, 1985.
Ferris, Timothy. *The Red Limit*. New York: Morrow, 1977.
Foust, Jeff, and Ron Lafon. *Astronomer's Computer Companion*. San Francisco: No Starch Press, 2000.
Gribbin, John. *White Holes*. New York: Delacortes, 1977.
Hawking, Stephen. *A Brief History of Time*. New York: Bantam, 1987.
Parker, Barry. *Cosmic Time Travel*. New York: Plenum, 1991.
Pickover, Clifford. *Black Holes: A Traveler's Guide*. New York: Wiley, 1996.
Shipman, Harry. *Black Holes, Quasars and the Universe*. Boston: Houghton Mifflin, 1976.
Thorne, Kip. *Black Holes and Time Warps*. New York: Norton, 1994.

9. TO THE ENDS OF THE UNIVERSE

Cornell, James, ed. *Bubbles, Voids, and Bumps in Time*. New York: Cambridge, 1989.
Ferris, Timothy. *Coming of Age in the Milky Way*. New York: Doubleday, 1988.
Layzer, David. *Constructing the Universe*. New York: Scientific American, 1988.
Parker, Barry. *The Vindication of the Big Bang*. New York: Plenum, 1993.
———. *Creation*. New York: Plenum, 1988.
Silk, Joseph. *The Big Bang*. New York: Freeman, 1980.
Smoot, George, and Keay Davidson. *Wrinkles in Time*. New York: Morrow, 1993.
Trefil, James. *The Moment of Creation*. New York: Scribners, 1983.
Wagoner, Robert, and Donald Goldsmith. *Cosmic Horizons*. Stanford: Stanford University Press, 1982.
Weinberg, Steven. *The First Three Minutes*. New York: Basic Books, 1977.

10. SEARCHING FOR THE ELUSIVE

Brian, Denis. *Einstein: A Life*. New York: Wiley, 1996.
Dukas, Helen, and Banesh Hoffmann. *Albert Einstein: The Human Side*. Princeton: Princeton University Press, 1979.

Fölsing, Albrecht. *Albert Einstein: A Biography.* New York: Viking, 1997.
Hoffmann, Banesh. *Albert Einstein: Creator and Rebel.* New York: Viking, 1972.

11. QUANTUM QUANDARY

Greene, Brian. *The Elegant Universe.* New York: Norton, 1999.
Gribbin, John. *In Search of Schrödinger's Cat.* New York: Bantam, 1984.
Pagels, Heinz. *The Cosmic Code.* New York: Simon and Schuster, 1982.
Zukav, Gary. *The Dancing Wu Li Masters.* New York: Morrow, 1979.

Index